R. H. Copperthwaite

**The Turf and the Racehorse**

Vol. 2

R. H. Copperthwaite

**The Turf and the Racehorse**
*Vol. 2*

ISBN/EAN: 9783337145903

Printed in Europe, USA, Canada, Australia, Japan

Cover: Foto ©berggeist007 / pixelio.de

More available books at **www.hansebooks.com**

THE

# TURF AND THE RACEHORSE

DESCRIBING

TRAINERS AND TRAINING, THE STUD-FARM,

THE SIRES AND BROOD-MARES OF
THE PAST AND PRESENT,

AND

HOW TO BREED AND REAR THE RACEHORSE.

BY

## R. H. COPPERTHWAITE.

Second Edition.

LONDON:

DAY AND SON, LIMITED,

LITHOGRAPHERS, PRINTERS, AND PUBLISHERS,

GATE STREET, LINCOLN'S INN FIELDS:

LATE DAY AND SON, LITHOGRAPHERS TO THE QUEEN.

1866.

# PREFACE.

In offering a few remarks to the sporting public upon the Turf and the Racehorse, I deem it at least necessary to render them in as simple a manner as possible, inasmuch as they are offered to the community at large.

Taking for granted that all the followers of turf pursuits, or lovers of horseflesh, are not Walkers, Johnsons, or Sheridans, it becomes necessary to write in language which can be plainly understood, instead of indulging in that high-flowing, flowery style, which tends more to test the faculties and bewilder the reader than to enlighten him on the subject ; and substituting what may be termed a superfluity of very fine English for instruction, thereby disguising the absence of practical knowledge: in fine, endeavouring to "spin out a long yarn" on a subject, with the merits of which they are but slightly acquainted. As the illustrious Moore said,—

> "Nine times out of ten, if his *title* is good,
>   The material *within* of small consequence is :
> Let him only write fine, and if not understood,
>   Why that's the concern of the reader, not his."

Others, preferring modern innovations, deal in poly-syllables, where perhaps monosyllables would be found more explicit, and to the point : for instance, now-a-days we read proofs such as the following. In an account of a good dinner we learn that "the tables groaned with all the delicacies of the season;" in returning from which, should a party happen to tumble into a ditch, we shall hear that "he became immersed in the liquid element." At Brighton, or any other watering-place, should a young lady while bathing happen to be drowned, the grievous intelligence is to the effect that, "having plunged fear-lessly into the bosom of Neptune, before the summer of her years had faded she sank into the silence of the grave."

It may perhaps appear presumptuous in me to attempt a small treatise on a subject which has been, and is so frequently, written upon by others—a subject, also, which is one of almost universal interest, and pecu-liarly calculated to challenge public attention : but having from boyhood owned horses, and studied their every movement, &c., and indeed I may add, occupied my mind with thoughts thereupon, when it might have been otherwise more beneficially employed—it is hardly to be wondered at, that as time wore on my passion, or taste for the animal, grew stronger, and as Horace says,—

" Quo semel est imbuta recens, servabit odorem
    Testa diu."

Or, in homely vernacular,—

"You may break, you may shatter, the vase as you will,
But the scent of the roses will hang round it still."

Whatever else I may have had to occupy my thoughts, there was one uppermost—"the Horse." Another circumstance prompted me to the attempt. In these sensation times, almost everybody seems to write something or other on this subject; and if my ideas or remarks do not coincide with the opinions of others, it can hardly be denied that, although "doctors differ," yet "two heads are better than one;" and it may so happen, that the reader will learn by a perusal of these few pages something foreign to his former ideas, and will then have what the late Lord George Bentinck was frequently heard to term "the best of the bargain," having the option of taking or rejecting them as he thinks proper.

There can be little doubt, that in speaking or writing on any subject which involves the interests of parties or professions, where differences of opinion must exist, the necessary consequence which may be expected to follow is, disapproval on the one side or the other, according as the doctrines of the writer may please or displease.

It too frequently happens that some hesitate to give candid expression to their sentiments, and adopt the sycophantic maxim of "running with the hare and holding with the hounds:" for even the great Cicero, when defending his friend Milo, feared to do so, but subsequent to the trial published a statement of what he had *intended* to say; upon reading which the latter exclaimed, "Oh ! if Cicero had thus spoken before my enemies, I would not now be eating figs in Marseilles !"

In presenting to the reader a few remarks, generally upon the subject of the Turf and the Racehorse, &c., they are rendered for the benefit of those who may think proper to accept them as useful; merely adding, that after about thirty years' experience I entertain sound reasons for my convictions, which are declared without prejudice, personality, or enmity.

I trust, therefore, that those who in their leisure hours may condescend to peruse them will make allowances for any errors which may creep in, or delusions under which the author may labour; and, as the poet said, remember that

> "Everything has faults; nor is 't unknown,
>   That harps and fiddles often lose their tone;
>   And wayward voices, at their owners' call,
>   With all their best endeavours only squall:
>   Dogs blink their covey, flints withhold the spark,
>   And double barrels (d—— them!) miss their mark."

# CONTENTS.

# TURF TOPICS.

"What reams of paper, floods of ink,
Do some men spoil, who never think!
And so, perhaps, you'll say of me;
In which some readers may agree.
Still I write on, and tell you why:
Nothing's so bad, you can't deny,
But may instruct or entertain,
Without the risk of giving pain."

ANCIENT history tells us that Nero loved his monkey, and Caligula his horse; indeed, to such an extent did the latter carry his affection for the animal, that he appears to have lavished upon him every luxury and comfort, to a degree exceeding (if possible) his barbarous treatment of his miserable subjects: his only wish, in the one instance, being that "the Roman people had but one head, that it might be struck off at one blow;" whilst in the other he was wont to swear by his *Incitatus*, whom he honoured with a palace, guards, and servants, and entertained at his own table, giving him gilded barley to eat and wine to drink in golden cups; clothing him in purple, with a collar of pearls; and on the eve of running his race having him carefully watched by a guard of honour, lest his rest should be disturbed. Is it, therefore, to be wondered at that this noble animal, possessing such silent

B

power and influence over the hardened heart and mind of
the cruel Caligula should, in the enlightened nineteenth
century, be the admiration of mankind; particularly when
we consider how far he tends towards our health, happi-
ness, and amusement, independently of his usefulness in
other respects?　But however beautiful and noble he appears under ordinary circumstances, and in other places,
nowhere does he *shine* so brilliantly, or show to such *perfection*, as when, in blooming health and condition, we find
him on the turf, ready to contend for victory; and how
gamely does the true thoroughbred struggle and strain
every nerve and muscle to that end !

It has been stated that the horse was the greatest
conquest ever made by man; and he has been, and is still,
found in his natural state in the deserts of Arabia and
Africa, and on the plains of Tartary, where droves of five
and six hundred have been seen at a time: but in Arabia
he appears in the greatest state of perfection.　The Arabs
love horses as their children, and live under the same
tents with them.　They surpass all other animals of the
desert in speed, and are so well trained that they *stop
as if shot* with the slightest touch of the rein; and
although the spur is unknown to them, they obey the
least movement of the foot.　As a further proof of the
sagacity of the animal, it is known that kind treatment
renders them so docile and fond of their masters that
they follow them about without being led.

The Arabs understand and are particular about the
pedigrees, which they divide into three classes: first, what
they term "first class," that is, "noble blood" on both
sides, which they can trace back for centuries; the
second, still "noble or ancient blood," but with a stain

on one side, which they term a " mis-alliance;" the third, " the common class." Those in a wild state are not so large as the domesticated; they are generally of a dark bay or brown, and their manes and tails much shorter.

In South America as many as ten or twelve thousand have been seen in a drove, and if by chance they happened to meet with a tame horse they have been known to surround him, and, by neighing and coaxing, endeavour to induce him to join their ranks and escape. Travellers have been left without means of proceeding on their journey in this way, being obliged to move in advance of their own animals in order to frighten away those droves. There can be no greater proof of the necessity of good care, and the benefits which must result therefrom, than the simple fact that the wilder these animals are the more diminutive they become, because they cannot, in their natural state, obtain the nourishment and comforts which they otherwise would; and, therefore, the British-bred horse bears a striking contrast to the Arab in size and bone.

For whatever purpose the thoroughbred horse may have been intended, there can be no question that he has been by man converted to a very useful one, and in the present day more so than ever; for, since the commencement of the present century, the numbers of horses turned to racing purposes have increased, to the present time, more than threefold, there being now more than three horses running in public contests for every one that ran in the commencement of the present century; an increase in numbers, which has been steady and regular, as well as enormous: for we find that, in 1802, but 536 horses ran, whereas there are at present from seventeen to eighteen hundred, or two thousand, according to the records of racing, contending annually for various prizes — a fact which proves that the "glorious pastime" has charms for many, whether founded upon pure love of sport or anxiety for pecuniary advantages.

What is the Turf? Let us take it from its very foundation, and it can hardly be looked upon in any other light than a bird's-eye representation of the world, with which no other pursuit, whether of pleasure, profit, or any nature whatever, can for one moment compare, in its representations of life and of mankind. Almost every true Briton appears to fancy it a duty, either from admiration or appreciation of the many enjoyments which it affords, or from curiosity, to visit the racecourse. Whatever may have been the ancient ideas on the subject of its pleasures, it can hardly be denied that it has fallen into hands, in modern times, which have turned it to purposes of business as well as recreation. If, according to the tastes and ideas of the present rising and enlightened generation, the turf were by possibility stripped of all its attractions, except that of witnessing a number of horses like a herd of buffaloes thundering over the prairies, the spectators would be few and far between, and the value of the beautiful thoroughbred would soon become seriously lessened : for, with all due respect to those who contend that certain high personages, who with kingly condescension grace the turf with their patronage and support, become proprietors of racehorses *solely* from love of sport, and with pure disregard of any profitable results, it is very questionable if even *they* continue to escape the electric influences of that metal which seems to possess such power over the human mind. It is, in fact, a field over which there is kept up a continued quest after the "universal idol;" the beauties, excitements, and pleasures of the chace varying according to circumstances. Some of the most zealous, although superbly mounted and well equipped, occasionally come to grief, through over-anxiety to excel all others; while, on the other hand, there is nothing more simple than by a cool and steady course to experience the enjoyments, and, at the same time, participate in the emoluments sought for. And it is very much to be regretted that fathers and mothers, old maiden aunts and rich uncles, should have heretofore formed such unfavourable and unchangeable prejudices against the turf; for it is far better the present gene-ration should benefit by a reasonable distribution of the coin of the realm than that it should be hoarded up for heartless, miserable specimens of humanity, who perchance would picture this glorious pastime to their heirs and successors as one to be for ever avoided. The turf is a pursuit sanctioned by Providence, to counteract the evil effects of the Satanic thirst of misers for worldly treasure, which now does, and always has, proved so dire and lamentable in its effects to their fellow-creatures, whose miseries and wants they calmly witness, while gloating over wealth which they prostituted their lives in accumulating, and yet do not enjoy, and which they would even cling to with tenacity, if it were possible to bargain for life with the king of terrors. Yet it is an ex-traordinary and glorious fact, and a gratifying consolation, that, in many cases, the successors of such detestable specimens of mankind are generally not only most liberal, if not extravagant, but are invariably staunch supporters and patrons of the turf; which, it must be confessed, affords the heirs or successors of such wretches ample opportunity of displaying their powers of distributing the cherished wealth of their ancestors, and thus preventing the possibility of the entire currency, which was intended for circulation, becoming buried, hidden, and concentrated in a few useless iron safes, where it could yield no good to anybody beyond the knowledge of its possession, which merely creates a grasping propensity for more.

Everything connected with the turf yields good to somebody; it is a wheel upon which fortune turns, casting benefits in all directions, in which all classes are participators. Its followers are invariably doing good one way or other, and are, and have at all times been, foremost to aid and assist their fellow-creatures, where their helping hand may have been required, and nevertheless they are, by a certain class of infatuated and prejudiced persons, the most abused body in existence; yet if the acts of some of those very parties, who, with the criticism and malignity of a Zoilus, censure the followers and patrons of the turf, were brought to light, probably it would appear, that while *they* had advanced in years they had not done so in virtue, however they may have differed in their selection of the course through which they may have elected to err. According to their picture of the turf and its followers, those who were ignorant of the real facts would almost be led to a belief that they were nothing short of a band of moss-troopers, with "*Vivitur ex rapto*" for their motto; whereas, in truth, it can boast that its foundation is based upon the solid support of the very picked pillars of the constitution, and that those pillars are propped up by adherents, the very soundest and most faultless in the British nation; so well tested, that any attempts to shake their strength would be as futile as gusts of wind against a tower of granite, or, as a certain learned Lord once remarked in speaking of O'Connell, " as useless as pelting paper pellets at the sides of a rhinoceros."

It is only very recently that an addition to the numerous proofs of the virtue and goodness of the turf's patrons was made manifest by one of its late most respected, lamented, and staunchest supporters, who bequeathed in charities alone no less a sum than thirty-six thousand pounds by his will, in which he forgot neither the unprotected widow, the helpless orphan, nor the faithful servant.

The man who feels disposed to err can do so in any pursuit in life, no matter what his object may be; but those who could condemn a noble and manly pastime because, forsooth, it may be accompanied by a wish or possibility to combine with its enjoyments that from which no earthly pursuit is totally free, must be possessed of minds not only capable of base acts, but prone to practise them if opportunity presented itself, or if the power to accomplish equalled the inclination to attempt. And it would be well if such people would recollect the words of Vousden's song,—

> " Let each man learn to know himself;
>     To gain that knowledge let him labour,
> Improve those failings in himself,
>     Which he condemns so in his neighbour."

Amongst the patrons of the turf rank the first men in the land, as well as many humble yet equally zealous. It would indeed be difficult to define exactly the difference in their objects, by taking or judging solely according to

rank or position. By many, who consider themselves
astute and competent judges, it is believed, that it always
was intended solely as an *amusement* for royalty and the
aristocracy, with permission to the community at large to
participate,—a condescension which the latter appear to
have availed themselves of beyond doubt, as some of the
most successful speculators are frequently very humble
patrons. It can hardly be denied, however, that the
great stimulus to the excitement and pleasures afforded
thereby consists in the anxiety of each, at least to prevent
his opponent from gaining the prize, whatever it may be, if
not to become possessed of it himself: perhaps this is the
mildest way of putting the case ; a most natural ambition
to take possession of any frail specimen of human nature,
no matter how exalted in society or however independent
he may be otherwise. It is said there are many who run
their horses solely for " honour and glory, and that sort of
thing : " it may be so, but it is more likely, and very much
to be apprehended, that if the ranks were confined to those
parties they would have little competition to apprehend,
and horseflesh would soon be at a discount.\* If money
could be obtained by asking for it (one of the last delusions under which any one is likely to labour, and the
very first to be relieved from), there would not be such
severely contested races, nor so many, as at present.
Ninety-nine out of every hundred persons who keep

\* How many men would, on the presentation of a check for
the amount of a Derby or St. Leger, request the Messrs. Wea-
therby to apply the amount towards the Lancashire distress, or
the London Hospitals ? For my part, I should back ' Current
Coin ' *versus* ' Glory ;' the former would be a tremendous favourite, and the result, doubtless, justify the confidence of its
backers.

horses would not *then* find as much pleasure, even in *winning* stakes, as they do *now* in the excitement of *trying*.

There lies the "kernel of the nut;" the rest is but the "shell," and the "fun" is in the breaking of it; and glorious fun it is, especially to those who require it most: for, after all, the pleasure or gratification cannot be so great to those who stand least in need of it, unless to one of those "money-worshippers," were he to make his appearance from behind his iron safe. There are plenty of good sportsmen in different positions, and there is nothing whatever inconsistent in one being a thorough sportsman, and, at the same time, anxious to benefit himself otherwise.

It is, however, not only amusing, but ridiculous, to hear the views some people form about men who are on the turf; and to hear persons who describe themselves as so, when they have never had any other description to give, and no earthly pretensions beyond, perhaps, having owned a fourth share in a plater. I once heard an intoxicated postilion at Bath, who had ridden home from the races, and was about being removed to strong quarters for the night, vow "he was *blowed* if he would go; that he was a racing man, and they *dar* not take him during the races."

So far as its amusements, there can be little question that the turf stands alone as a pastime—that it reigns supreme above any other. Even the fox-hunter could barely live without its addition; but the question is, for whom is it fit? or to whom is it suited, as far as becoming proprietor of a stud, with its necessarily heavy require-ments?—questions best answered by those who may have had the pleasing gratification of trying their hands. To the young, ardent beginner, who may have suddenly fallen into the receipt of a fine fortune, there is no possible arena wherein he can better display all his powers, or carry out his wishes as to investing his capital, and proving himself a worthy or liberal deputy for the distribution of a long-cherished treasure. Yet it by no means follows that he cannot have, to his heart's content, unbounded amusement without a farthing loss; nay, even with plenty of profit, provided he keeps within bounds and acts with reasonable judgment and prudence. If he does not himself possess sufficient judgment, there are plenty capable of teaching him, who will be delighted to instruct him, and give him the benefit of their own perhaps dearly-bought experience, which is generally the best of all. His great care should be, while endeavouring to select good horses, not to neglect or forget the more important point of making a good selection of his friends and advisers, and to take heed lest he should fall in with those who, in endeavouring to regain some of their own " experience-money," might charge too high a price for their instruction : for, unfortunately, it must be confessed, that the natural love of self is not forgotten on the turf, any more than elsewhere, whilst it affords the amplest opportunities of gratifying such a failing, and on the grandest scale. And not only beginners, but even old heads, have sometimes more to fear from those nearest to them in whom they may have, in over-confiding weakness, placed implicit confidence, than from their opponents, for the common laws of nature dictate that in such pursuits, as in every other in life, love of self should reign predominant in the human breast.

Still there are plenty, whose upright and straightforward disinterestedness would not permit any unworthy motives to interfere with their good intentions to benefit their younger friends by their advice; but it is from neglect of caution in selecting such monitors that so many young men have heretofore been led astray, and sadly victimised: it is, however, a happy reflection that the march of intellect, in the present day, is such as to leave little need of apprehension in the minds of their well-wishers, for the rising generation appear to be very competent to take care of their own interests, and, like the young horses of the present day, are showing a marked superiority in that respect, when contrasted with those of former days, who have "broken down" in a very short time; indeed, in racing parlance, without ever having developed or displayed much form.

Then, assuming that the reader may be disposed to enter upon the stage, and try his fortune, I shall take the liberty of supposing him a novice, and offer to him any little information which may be within my power towards his enlightenment on the subject; and will suppose that he is about to commence "a nice little establishment," for the purposes of pleasure combined with the probability of success, and with a dash of honour and glory.

It appears to me, that the man who ventures upon a breeding establishment with any view beyond mere amusement, has frequently more trouble, expense, and risk before him, than he may fancy at first thought. The paddocks may have their charms for the eyes of the casual visitor, who may admire a fine old mare, perhaps the winner of the Oaks or the Thousand-guineas stakes, with a promising foal by her side, from which the owner expects even greater success : he may fancy he is looking at the winner of a future Derby or St. Leger while gazing on a promising yearling; or may visit the box of a stallion, probably the winner of both, yet doomed never to get a winner of either—which is no uncommon occurrence. Those are very agreeable visions, no doubt, and, as far as the *pleasure* of the speculation is concerned, afford it in abundance, provided it is a matter of little concern to the breeder whether they prove otherwise profitable, or that such is a secondary consideration ; yet it is a speculation surrounded with perpetual torments, anxicty, and losses : the latter frequently on so large a scale, that it is very questionable whether a party using racehorses for profit would not, in the present day, find a more beneficial and economical mode of keeping himself supplied : for, no matter how careful or intelligent the stud-groom may be, while grass grows, or water runs, he will now and then have to announce to his employer something or other, in the course of his duties, which will have upon the latter any effect but one tending to increase his appetite, or improve his digestion. If you pay a large price for a sire, particularly an untried one, no matter what his perfections or qualities may be, you may in vain make use of all the gift of speech and persuasion with which Providence may have endowed you, to induce people to believe that his shapes, blood, and so forth, are what they ought to be. You hear then of the kind (?) remarks of a neighbour, who may happen to be proprietor of a rival stallion, that yours is either a "roarer" (according to reports, almost every stallion is a roarer—a most mistaken idea, elsewhere explained), or an uncertain foal-getter (a very likely matter, for reasons also explained), or some such observations. When I purchased ' Mountain Deer,' and imported him to Ireland (where it is said horses are so fast deteriorating of late; and little wonder, although there are good sires enough), my groom used to report to me the various opinions of parties who came to see the horse, which certainly were about as flattering as they have since proved to be valuable.

" Has anybody been to see the horse?"

" Yes, sir: Mr. So-and-So, and some other gentlemen."

" What did they think of him?"

" Not much, sir. One said he had flat sides; another, that he had bad legs; and a third, that he did not like his white face. I *tould* them, the *divil a pinsworth they knew about it—that he cost too much not to be good.* But there was one gentleman said he liked him very much, and that he would send three mares for half price, if he got the keep of one of them gratis."

A breeder must have patience and a long purse, for there must always be great wear and tear of capital, in purchasing untried stallions or mares, which, until their characters at stud have been established, are of little profit, and frequently turn out worthless. Then, again, the losses which are experienced through death or accidents, the missing of mares, and, in short, the continual drain upon the exchequer, render it a most hazardous speculation; in most cases a purchaser of stock adopts by far the better and more economical course by attending public sales, or still better, by purchasing privately from parties, who perhaps are much more easily dealt with, and from whom bargains are more frequently obtained than at fashionable auctions, where competition is often so spirited and sometimes so very *hot*. But if any man attempts to breed for sale in the present day, and does so from any but the best, most fashionable, and running blood, both on the sides of sire and dam, he might as well, and much better, present his money to some charitable institution.

Then, when the sale-day does arrive, the purchaser will best consult his own interests by obtaining the best lot, even at the top price; not that it is impossible, or even improbable, that the very highest priced one might turn out the worst, or *vice versâ:* yet, as a general rule, speculators in horseflesh have become such masters of their business that they generally "hit the right nail on the head," the big money frequently succeeding, partly from the lot being competed for by experienced judges; although curious exceptions are frequently seen, where valuable animals are sold, and even forced upon purchasers, for merely trifling sums, both at public auction and by private sale. The following was an extraordinary instance. There is at this moment in Her Majesty's stud at Hampton-court paddocks, an animal named 'The Deformed;' without exaggeration, as magnificent a specimen of a thorough-bred mare as any to be found, and well worth the time of any lover of horseflesh to look at. Her size, symmetry, and blood-like appearance, together with substance, almost defy comparison. She is by 'Burgundy' or 'Harkaway,' dam 'Welfare,' by 'Priam.' I purchased this mare, when a yearling, for 15*l*.; she being at the time engaged in four large stakes, all of which she won, besides many others, both in England and Ireland, and subsequently ended her racing career by winning Prince Demidoff's cup in Italy.

I sold her to Captain Scott for 1500 guineas, repurchased her as a brood mare for 300, and subsequently sold her to the late Marquis of Waterford for 600 guineas, at whose sale she was purchased for Her Majesty. She was thus named by me from the fact that she turns her left foot rather inwards (her half-sister, ' Mag on the Wing,' once in my possession, did so likewise), and walks and gallops with a peculiar, round, wide-sweeping action, like old 'Harkaway.' She has the temper of a lamb, the propelling power of a steam-engine, the eye of a gazelle—in every shape defying exaggeration from the pencil of a Herring or a Hall; and has proved herself, both in England and Ireland, an extraordinarily good mare.

*The* Marquis's name calls up recollection of an instance of the ill-luck which, during one week's racing at the Curragh, attended that much-respected and deeply-lamented nobleman; and how good-humouredly he bore with it! He had sixteen horses running during the meeting, and did not win one race, although in some he ran two or three. The races over he commenced laughing heartily at the idea, and there being a travelling show opposite the stand invited a certain popular Baronet, a particular friend, to accompany him thereto, forgetting all the disappointments of the race-week. They repaired to the exhibition, and finding amongst the " curiosities" two pelicans, he challenged his friend to match one against the other for a small wager. The match was made, the birds ran for the fish, and the Marquis won by a " bill ;" at which success he appeared as much elated as if he had won a Derby. Well was he named, and long will he be known and remembered, and his memory respected, as "*The* Marquis." We ne'er shall see his like again !

In the formation of a stud for racing purposes, a great deal must depend upon the intentions or wishes of the party, as well as the length of his purse; a very great mistake, and one frequently made, being, that of purchasing *too many,* sometimes of a moderate stamp, instead of confining the number to *fewer* of *first-class promise* and *quality,* for the expense of the one is as great as that of the other. The principal question for a beginner is, as to the best way to accomplish his wishes and suit himself. Then, suppose his object be to obtain the very best class, regardless of price, having in view the "Blue Riband," Doncaster and Goodwood cups, &c., and that he is one of the " honour-and-glory " party, his simple course must be to attend the sales where such are advertised for public auction, and seek (if he be not himself a competent judge) the assistance of those who are : but then he must make up his mind, as a general rule, to pay high prices, for in the present day the competition for yearlings of promise is very great indeed, and the prices exceedingly high, especially at fashionable auctions, where, as a matter of course, such animals are most likely to be found : for it is wonderful the very high prices which some breeders pay when purchasing brood mares, sires, &c. Such auctions are always attended by parties, either upon their own behalf or in the interest of others, whose judgment might, figuratively speaking, be compared to that of Sylla, who declared he saw "many a Marius in the stripling Cæsar." Still, the fluctuations in prices, the competition and judgment displayed, are extraordinary, and only equalled by the contrast between the want of judgment of some and the soundness of that of others ; the inexperienced employer in some cases being encouraged, by the sagacity of an over-anxious trainer, to become a purchaser at all hazards, and at any price, rather than let the stable be empty.

The prudence of paying very large prices for yearlings is a question admitting of difference of opinion, and depending very much upon circumstances; and to arrive at a conclusion as to what the real value of a first-class yearling—say an own brother or sister to a celebrated horse — may be, is a most difficult task, and one which becomes so even to the seller. It is truly said "that the value of anything is what it will bring," and it is equally true that the fairest and most simple way to ascertain that value is by public auction; still, the question of *prudence* in always following the rule, in the case of such property as that in question, admits of doubt, for the following reasons. The prices realised at auctions, for some lots, are more artificial than really consistent with reason, because two or more anxious wealthy competitors may make up their minds " *to have the animal,*" regardless of expense — such resolutions being frequently formed, not so much upon the actual shapes or qualities, as upon the fact that they had been successful with the same family, such as an own or half-brother or sister — consequently, very much to the satisfaction of the seller, and sometimes the astonishment and amusement of the bystanders, who may happen to be really sound judges, the animal is sold at probably three times its real value, without perhaps a bid from others, who had more judgment, and as much means. Then comes another lot, in reality worth more, yet, not having the *prestige* referred to, the competition is not so spirited.

These remarks are merely applicable to the *prudence* of paying *extravagant* prices for yearlings, *because others*

c

of the family may have done good service for *parties individually.*

But taking into consideration the great risk and lottery in young stock, and how frequently the most promising turn out the very worst; considering the matter in a mercantile point of view, the prices sometimes paid are fabulous, and how frequently do such " swans " turn out inferior to "geese !" The 1800 guineas 'Lord of the Hills,' ' Voivode,' &c., for example. The prices of young stock must be viewed as a question of fancy, which can be indulged in according to the length of certain purses, as well as taste. It is but common reason to suppose that he who gives a thousand pounds for a yearling ought to, and in most cases will, have a better chance of a racehorse than a purchaser at fifty pounds : still, curious " turns-up " occur, where animals are purchased for very small sums. I once purchased a St. Leger winner for less than fifty pounds, and afterwards handed him over to a friend. Many of the best horses have been bought, both at auction and at private sales, for moderate prices : such as ' Thormanby' at 350 guineas, ' Voltigeur' at less, ' Caractacus ' less than 300 guineas, ' Kettledrum ' at 350 guineas. ' Early Bird ' cost but 70 guineas, besides many others, at prices varying from 70 to 200 and 300 guineas. 350 guineas seems to be a *fortunate* price : ' Chattanooga,' winner of the ' Criterion,' cost that price. Many of the best horses have been picked up quietly for very small prices by private sale, which is by far a more prudent way of purchasing, for many reasons ; and there are few owners who will not sell when offered a fair price. But still, some of the greatest bargains are had at public sales ; where the breeder has perhaps, in his private

calculation, booked one as likely to bring 500 guineas, he
has been knocked down perhaps at 100: so much depending on the whims or fancies of purchasers. In fact
the prices, like breeding, and everything connected with
horseflesh, are a lottery; and if a breeder is occasionally
remunerated with a fancy price he is deserving of it, for in
the long run, when matters are wound up, he requires a
" lift " to square his account : and some are richly deserving
of encouragement, from the spirited and liberal manner in
which they purchase when forming their breeding establishments: not that they are likely to add one guinea to the
" tot," by reason of any generous impulse on the part of
the bidder, beyond what the qualities or merits of the lot
may justify; for if they relied one iota on that slender
thread they would find any other mode of investing their
capital far more profitable. However promising an own
brother to a 'Flying Dutchman' or a 'Blair Athol' might
be, should he march into a ring with a blemish or questionable formation, the fall of Messrs. Tattersall's hammer,
notwithstanding their persuasive eloquence, would soon
announce a figure anything but encouraging.

I remember upon one occasion asking Mr. Tattersall what he fancied a certain colt, of a number about
to be sold at his establishment, would realise ? His
reply was something to the effect, that " I might as
well ask how many stars there were in the sky, so
much depended upon the intentions of parties to purchase; that price to some was a secondary consideration,
if the animal suited." Purchasers must frequently cut
their coat according to their cloth ; but my advice to any
reader would be to avoid over-excited competition, and
if not sufficient judge himself to obtain the assistance of

parties who really are, who have proved that they are so, and not pay attention to persons who would almost talk one out of his senses about shapes, and some of whom have spent half their lives breeding, buying, and racing, and yet whose attempts to produce a racehorse remind one of a man endeavouring to " open an oyster with a pitchfork." Considering the number of young ones purchased by some parties, and the sporting prices paid, it is astonishing how few good animals they select. Others purchase most dreadful samples, and having once done so, even cling to them with the tenacity of a " sinking sailor clinging to a mast," instead of taking moral courage, turning them loose, and letting whoever catches them keep them. The first loss is always the best in such cases; and it is to be presumed that some day a purchaser will hit upon a prize.

Although in most pursuits in life the old motto, " *Experientia docet,*" is verified, the case of judgment in horseflesh, to a very great extent, is, in my opinion, an exception; for I believe that all the practice and experience that could possibly be bestowed on some would not make them judges: in short, that there is a taste necessary; and that some have naturally " an eye for a horse," while others, as a certain gallant old general would say, " hardly know a horse from a hen," can scarcely be questioned. How many men, having spent their fortunes and lives in this pursuit, have gone *under the turf* without having won any prize worth mentioning? Again, how many still live, who are day after day paying high prices, and spending large fortunes, with the same result? But it is not always want of judgment or means, but frequently mismanagement and want of the care that becomes necessary when the owner finds the young racehorse promising; many a valuable young one is, and has been, rendered worthless, through the negligence or ignorance of those who are paid to attend to its requirements.

It has been remarked with truth, that "horses run in all shapes," yet it cannot be denied that they do so *more frequently*, and *better*, when those shapes are *good*, and of the approved and tried formation; and it frequently happens, that the very animals that many *connoisseurs* call ill-shaped or *ugly* are quite the contrary : for one possessing the best points in the world, where those *good points* are *required* and *put together* as *machinery* should be, in the *proper place*, may, at first sight to a novice, or even to one who *fancies* himself a judge, appear an *ugly* horse, and *really be one;* but some of the *ugliest* horses are the *very best shaped* when properly looked over, and there alone rests the point where the natural eye discovers the probability of the animal turning out well. Who would have picked out such a horse as 'Fisherman,' if he were inexperienced? or who would have chosen 'King of Oude?' the latter, perhaps, the most extraordinary-looking animal that ever was foaled; yet, when looked over, they were combinations of magnificent racing shapes. The *worst-shaped* horses, *for racing purposes*, are frequently the *handsome* ones; and it would be hardly going too far to add, that they are invariably the *very worst*. The greatest failures are generally very *handsome*, and only fit for Rotten Row, where they are most valuable, no doubt. The *machinery, properly put together*, like in any railway engine, is the point; the shapes, taken *separately*, may be *perfection*, but if not *properly screwed together* in the particular or principal place, they are as useless as the works of a watch with a broken spring. And the principal point which calls for the attention of any person is the *coupling* of the *two ends;* and the chief, if not almost the entire success, depends upon the *propelling power* from the *hind quarters,* which should be well placed and sufficiently turned under. 'Leamington,' probably, would afford about the best example of any horse that could be named, and would certainly be a very fine model for any novice to learn from in other respects.*

But, no matter how great the experience or judgment may be in purchasing horses at a year old, or untried, it is a great lottery, for some of the best-looking have not been *foaled* with the *gift of speed,* and no training or time can give it where Nature has denied it; however, they will give stamina, and add staying powers : even an overgrown or large young one must show a *turn* of *speed,* to a certain extent. The chance and lottery in yearlings are wonderful, and the greatest disappointments are experienced from the best-looking : as Shakespeare says,—

> " Oft expectation fails, and most oft there
> Where most it promises ; and oft it hits
> Where hope is coldest and despair most sits."

Although the prudent course is invariably to adhere to the *own brother* or *sister* of good horses, or the produce of *tried brood mares* and *stallions,* yet in the case of the

---

* The shapes of a racehorse may separately be perfection ; a man may know what good shapes are, and yet be just as far from knowing what a racehorse is as a schoolboy, when he tries to put a toy map of Europe together, which he had never seen before.

produce of a mare that had perhaps highly distinguished herself on the turf, but not been *tried* at *stud*, such produce being put to action would, if good-looking, realise a high price. Then suppose another yearling, the produce of a daughter of that mare (never, perhaps, having been trained) by the same sire, equally good-looking, and in other respects desirable, the one would in all probability bring double or treble the price of the other; yet, to my mind, the chances would be in favour of the produce of the young untrained mare : that is to say, provided in appearance, shapes, size, &c. such produce equalled that of the old mare, and that both had in every respect "equal main and chance," and the earlier the trial was made the more I would rely on the young mare's produce and yet at auctions the habit generally is to run after the other, although we have innumerable proofs against it. My principal reason for arriving at this conclusion is, that we have every day proof of first-class mares having been *long* in training, and having done *severe* work, highly distinguishing themselves when put to *stud*, and yet frequently proving perfect failures; whereas some of the best brood mares have been useless as racehorses from various causes; and if the young fresh mares more frequently had the chances of old ones, they would be even more successful. But owners do not generally give them that chance, *because* they are disheartened from the fact that they have not distinguished themselves; whereas they are, on the other hand, carried away to expect wonders from the old brilliant performer.

A reference to the performances of the produce of the very best mares — say the Oaks winners, &c. — will prove that they have been as a lot perfect failures; to which fact I

shall more fully refer under the head of " Brood Mares :"
another remarkable fact being, that the *early* produce of
such *distinguished* mares, even though *subsequently first-
class*, are not the best. Take old 'Beeswing,' for instance; one would have expected the best vintage from
her alliance with 'Sir Hercules,' yet although the produce was 'Old Port,' it was not good. But, to my mind,
the most prudent class of speculators are those who pa-
tiently await the opportunities (which they seldom have
to wait long for) of purchasing horses, say at the end
of their two years, after they have run and shown *some*
form, perhaps overgrown, but of an improving sort, and
from staying strains. Such a course, in the first place,
has one advantage, that it gets rid of the risk to a
very great extent, or rather the lottery; for there one
buys with his eyes open, and it is almost incredible
the state in which some horses are brought out to run,
some as fat as pigs, but more frequently galloped to
death, leg-weary, and with skins as fast as the bark on
trees, and frequently parted with for trifling causes, not
known or understood by their owners. Taking into consideration the fact that the largest stakes are realised by
horses of moderate pretensions in handicaps, wherein
they defeat the largest priced yearlings, from which they
receive as much as two stones even at three years old;
and, moreover, when we even find old horses literally
turned loose; it is high time for speculators to open their
eyes when, without throwing sand in those of the handicapper, they can win the richest stakes run for. Not to
speak of three-years-old alone, we sometimes find owners
display their ability and judgment by winning some of
the largest races with old horses at weights, which would

lead one to fancy that it will shortly be necessary to train monkeys for the pig-skin ; while it is impossible to say at what weights first-class three-years-old will, in some years hence, be visited.

To use racing phraseology, it would appear that " Handicapping" is the father, " Money" the mother, and " Pull'em" and " Scratch'em" the children of the sins against the Turf. The enemies of the paternal parent (the maternal one has none) avow that he is instrumental in ruining the breed of horses in general, and punishing good ones in particular ; that he holds out inducements and temptations, tending to soil the morality of the turf, by testing the integrity of its patrons to too great an extent, as to their regard for the maternal parent, and frequently obliging them to sin against the rules or their code of laws and honour.

If it were possible to do away with the system altogether, much benefit might ensue ; for, notwithstanding the vigilance of the most scrutinising adjudicator of weights, " Will-o'-the-Wisps " will occasionally appear visible for a moment, but speedily became extinguished by lights still more brilliant.

If the object of those in authority really be to abolish the practice, probably the most desirable course, as well as the one most likely to have the desired effect, would be to muster up courage, and concentrate their forces ; and without favour or affection, regardless of rank or position, deal out their judgment with an even and impartial hand : and thus render their warnings and decisions more to be respected and dreaded, and prevent the possibility of any transgressor saying with the Scythian, " that laws were cobwebs, wherein small flies got caught and larger ones broke through;" and assimilating to himself the position assumed by Dionides the pirate with Alexander the Great, who, when asked "how he dared to trespass on his seas?" replied, "that he did it *for his own profit, as Alexander did himself. But,*" added he, "*I, sire, who rob with a simple galley, am called a pirate; but you, who plunder with a great army, are called a king!*"

The sacrifice of a shipload of poor pirates like " Dio," would not have half so beneficial or salutary an effect as the offering up of one aristocratic holocaust, if he were caught playing "Will-o'-the-Wisp," or exceeding the bounds of decorum in his poaching propensities, after either glory or gold. For if such an extraordinary phenomenon were to be presented, the effect upon the astonished multitude would be so great, that probably a system so baneful to the morality of the turf would be for ever abolished; a consummation, however, which, while handicapping exists, it is much to be apprehended will not be accomplished until about that period when the industrious children of Israel return to Palestine.

Probably those mystifying "Will-o'-the-Wisps" practice their errors, and justify them on the principle, that "when they go to Rome they do as Romans do;" that, as others in their struggle for the prize resort to such stratagems, *they* cannot "tie one hand behind their back," but must fight them with their own weapons; and may have the temerity to add, that if they did not do so they would have no chance, and be totally debarred the possibility of success: their opponents would

"Like lions o'er the jackal sway,
By springing dauntless on the prey."

The duties and responsibilities of a handicapper are at all times most arduous, and sometimes thankless; requiring not only a thorough knowledge of the numerous animals (sometimes of men, which perhaps should not be taken into consideration, and too frequently is in various ways), but also skill and experience generally in matters of horse-racing: in short, a thorough knowledge of the horse himself, as to condition and otherwise. Even supposing that all the horses submitted to him had previously displayed their true merits, it is a task of no ordinary undertaking, but requires the brain of a Chancellor of the Exchequer. But when the conscientious adjudicator, anxious to measure out justice to all alike, finds that he is dealing, not with facts but shadows it must, indeed, especially to an upright and conscientious man, be anything but agreeable or encouraging.

Fancy, after all the pains and trouble consequent upon the framing of an immense handicap, some great, fine, four or five years old, coming bounding in like a deer before a pack of hounds! Why one, in the amazement and excitement of the moment, hardly knows whether to condole with the chagrined adjudicator or laugh with the artful fox, the destroyer of his hopes; while the latter chuckles over his success, to the mortification of many who may have been overmatched at the same game, and found themselves out in their own private calculations.

Admitting that the duties of the handicapper are most onerous, and that the subject is one which involves serious amounts, and is, therefore, of vital importance to the racing public, especially to the owners of horses, it must be deserving of consideration what those duties are, how far they extend, and where they terminate; a very nice point indeed: taking for granted, that they have for their object a fair and impartial adjudication of weights; but according to what? Here is the question, the main point, and the stumbling-block.

Let us suppose that a given number of horses are entered of various ages—or even of one age, for simplicity's sake; that all of those animals have run in public and displayed certain form, and that the handicapper be a competent one. To what end is he justified to carry his inquiries? upon what grounds is he supposed to found his judgment, or form his opinions of their respective merits, beyond the manner in which they *absolutely performed?* If the difficulty ended with arriving at calculations exclusively upon *that head,* and if it were the case that all had fairly tried their best on the several occasions of their running, the difficulty would be but slight. But suppose that various stratagems are resorted to for the purpose of misleading or misrepresenting their merits, such as want of condition, running them out of their natural courses, or the old-fashioned go-the-whole-hog system, which, when done in a slovenly manner, is sometimes discovered, and occasionally punished, either by a subsequent "crusher," or if "guilty" pleaded, notice to quit *in toto ;*—how far is a handicapper justified in forming surmises or coming to conclusions, in some of which he may be sadly mistaken? Is he to listen to whispers from idle prattlers, who, perhaps, with the assistance of a glass in their eye, may have discovered a "mare's nest," or fancy they saw something which never happened? or is he to attend to the innuendoes of interested parties, who from either personal interest, private *pique*, or jealousy, may throw out hints, if not assert deliberate falsehoods, to the prejudice not only of owners, but perhaps others? It is quite natural, and consistent with reason, that he should form his opinions upon the running of horses, taking into consideration their condition and all the points in relation to their true merits, which his experience may dictate; but any man to be a handicapper must be a thorough judge of the animal, his condition, his probable improvement, and all such matters. The fact is, that if people are disposed to mystify, it is a task much easier to accomplish than for a non-subscriber, however high his position, to effect an entrance into the subscription-room at the " Corner," without the knowledge of the Argus-eyed and indefatigable overseer thereof. The handicapper is not in a position to *assert* that which he cannot *prove*, however he may *fancy* it. " Sir, you are a ' Will-o'-the-Wisp !' I 've got you ! and *I* shall hold you until *we* tell you to vanish." But when an owner, who may not have resorted to deception, has run a wretched animal good for little more than consuming corn, and finds himself politely treated to perhaps the top weight, what must his astonishment be? Therefore, however unpleasant it may be to such owner to brood over the misplaced or unmerited compliment, still the handicapper has no other alternative or mode of giving the " gentle hint," that he entertains a higher estimate of the animal's pretensions than of the owner's straightforwardness, which conclusion may have been formed through mere idle report. He is thus led to act upon reputation, not upon character — two very different things.

As long as handicapping exists, it will be surrounded with grumbling, vexation, and discontent, on the one side or other; and the position of the adjudicator will be any-thing but an enviable one, however it may tend to keep the brain in exercise. It rather resembles that of the old man, his son, and the ass, in the fable—he pleases nobody; and might be compared, in some respects, to the description I once read of a crown,—

> " 'Tis to bear the miseries of a people,
>   Hear their murmurs, feel their discontents,
>   And sink beneath a load of care;
>   Have your best success ascribed to fortune,
>   Fortune's failings all ascribed to you."

It is to be presumed, that, so long as it exists, the object of a handicap will be to give an equal chance to animals of various ages, according to merit. Then it appears strange that such a cry should be so very frequently raised, if a four or five or aged horse wins —an argument, in fact, that they should be debarred from a possibility of defeating young horses; whereas every day is adding fresh proof that the young ones can almost do anything with old ones, especially at the end of their three-year-old year, when they are on the very verge of four years old. Take the Chester cup, early in May, and what do three-years-old not accomplish? Take the Cæsarewitch for years, and even so far back as 'Faugh-a-Ballagh,' with his 8 st.; the 'Baron,' 7 st. 8 lbs., &c. The changes and improvements in three-year-olds of late years, as well as the marked alterations in handi-capping, are wonderful. For example,—

### Chester Cup, 1842.

| Lancrcost | aged | . | . | 9st. | 9lbs. |
|---|---|---|---|---|---|
| Cruiskeen | „ | . | . | 8 | 0 |
| Retriever | 6 years | . | . | 7 | 8 |
| Millepede | 4 „ | . | . | 6 | 10 |
| Alice Hawthorne | 4 „ | . | . | 6 | 0 |
| Topsail | | | | | |
| Tripola | 3 „ | . | . | feathers | |
| Proof Print | | | | | |

In the above 'Lanercost' accepted, and it can hardly be questioned that 'Alice Hawthorne' "looked well in."

### Chester Cup, 1844.

| Alice Hawthorne, 6 years | . | . | 9st. | 8lbs. |
|---|---|---|---|---|
| Retriever | aged | . | . | 9 | 2 |
| St. Lawrence | „ | . | . | 9 | 0 |
| Gasparoni | „ | . | . | 5 | 9 |
| Red Deer | 3 years | . | . | 4 | 0 |

In those days, it was looked upon as an act of absurdity, even to "enter" a three-years-old for the Chester cup, heavy as the weights were set, until the late Lord George Bentinck astonished the talent by winning the above race with 'Red Deer,' ridden by that extraordinary little jockey Kitchener. The accuracy of his lordship's judgment has been wonderfully borne out of late years; for when we refer to the weights, with which the three-years-old contend now-a-days, it is enough to make people ask, or wonder, what they will carry and win with in 1880?

Contrast the foregoing with the following, in which horses of all ages contended, and 'Stampedo,' five years old, winner of the great Northamptonshire stakes, without

his ten-pound penalty, headed the list and accepted with
8 st. 5 lbs. : —

### Chester Cup, 1862.

| | | | | | | |
|---|---|---|---|---|---|---|
| Tim Whiffler, | 3 years | . | . | . | 6 st. 11 lbs. | 1st. |
| Investment | 3 „ | . | . | . | 6    8 | 2nd. |
| Brighton | 3 „ | . | . | . | 6    0 | 3rd. |
| Sappho | 3 „ | . | . | . | 6    5 | 4th. |

Here are four three-years-old, the first four beating
'Caller-Ou,' 4 yrs., 8 st. 11 lbs.; 'Fairwater,' 4 yrs., 8 st.
7 lbs., &c.   The reason why three-years-old have been
and are becoming more frequently successful, and old
horses are diminishing so much during the present
century, may be gathered from the returns, furnishing
the following results.

In 1802 the two-years-old were about one-seven-
teenth; in 1860 they were about one-third.   The three-
years-old, in 1802, were one-fourth; in 1860 they were
one-third.   The four-years-old, in 1802, were one-fifth; in
1860 they were one-fourth.   The five-years-old, in 1802,
were more than one-half the entire number; in 1860, but
one-sixth of the entire were five or upwards, and but one-
third of all horses were four or upwards.   The increase
in the number of horses of all ages running from 1802
up to the present time, was, in 1802, 536; in the present
day, nearly 2500.   In 1802, but 100 three-years-old ran;
in the present day there are about 600 contending.   About
30 two-years-old ran in 1802; in the present day there
are about 700; whereas in 1802, 280 of five and
upwards started, and now there are not more than about
the same number.   There are at present about 1500
races, contended for by all ages, of the following distances: —

Half-a-mile and under . . . . 256
Over half and under a mile . . . 474
One mile . . . . . . . 281
Over a mile and under two . . . 301
Two miles and under three . . . 177
Three miles and upwards . . . . 25

As far as the Cæsarewitch is concerned, of late years, three-years-old have shown what they can accomplish. ‘Dulcibella’ literally could have trotted in with 6 st. 11 lbs.; she won by ten lengths in a canter, and four-years-old in with 6 st.; ‘Asteroid’ was third, with 7 st. 6 lbs., beaten a neck, for second; and in the present year ‘Ackworth’ was third, with 7 st. Had such horses as ‘Faugh-a-Ballagh,’ ‘The Baron,’ or ‘Asteroid,’ been in with 7 st., where would those two mares, ‘Thalestris’ and ‘Gratitude’ (neither of which had previously shown even moderate form), have been? It is quite certain *something must* win, and it appears strange, when two four-years-old run so well-contested a race to a head, and a three-years-old third, with 1 st. less than horses of his own age had won with, that there should be any reason to complain as far as the first rank were concerned; and it should be borne in mind that, in the Cæsarewitch especially, all horses do not finish out, and many run merely for a certain distance, as a trial for the Cambridgeshire. A handicap is nothing more than a handicap — they cannot all win; and it is equally certain that their owners will never be all satisfied, for the world is a grumbling world; and the Turf, although a world of wonders in itself, is not a likely one to be the exception. The great mistake that has been made for years, was not appreciating what young horses could really do. In the Liverpool cup it was considered a regular "settler" to visit a three-years-old with 6 st. 7 lbs., for not one of

D

them could win with a weight of the kind, until 'Charles the Twelfth' did succeed: he was a first-class horse, and won the Doncaster St. Leger. About the same period we found old horses handicapped at 10st. 6lb., giving 5st. to three-years-old in the Goodwood stakes, at that advanced season of the year; whereas, now-a-days, the young ones can do almost anything with old horses, at very little difference of weight. For instance, 'Pretty Boy,' three years old, as well as I remember, won with 7st. 8lbs., and 'Elcho,' carrying within 1lb. of 6st. defeated 'Starke,' six years old, at a difference of 24lbs. in the Goodwood stakes. 'Starke,' same meeting, won the Goodwood cup, carrying 8st. 10lbs., defeating 'Thormanby' and the 'Wizard,' first and second for the Derby, and a good field of horses of various ages. The fact is, many three-years-old, at the end of the year, arrive at their best form, especially those that may have been trained and ran early; for it is "the pace that kills," and what some gain in stamina by age they lose in speed.

In the late Cambridgeshire the success of the winner 'Ackworth,' with 7st., appears to be looked upon by most of the *cognoscenti* as a great performance. No doubt it was a *good* one. But they appear to forget what several others did previously, which, when contrasted with it, will hardly tend to prove that there was anything *wonderful* in it: nor does the result, as far as the race is concerned, in other respects prove it so. It must not be forgotten that the distance was shorter, although a more severe course, than the Cæsarewitch; and it certainly appears strange, the young horse having seven pounds less weight than in the longer race with 'Gratitude,' the four-years-old, especially further on in the season. But how many other three-

years-old have won, and only been just defeated by a
head, with far more weight than the late winner?—' Odd
Trick,' ' Saunterer,' &c.   Moreover, it is an admitted and
well-proved fact, that horses can give away to each other
more weight at two years old than at any other age; and
there can be no reason why the same rule, to a certain
extent, should not prove, and in fact it does prove, pro-
portionally the case with the three-years-old.   The
difference appears greater upon paper than it really is,
when horses of a superior class are put alongside moderate
ones; for it is wonderful what really good young ones
can do with those of a different age, as well as those of
their own, especially where there is a great turn of speed :
for horses deficient therein are always in trouble, and
many of them so much so that, in fact, they could not *win*
with any weight.   Many persons who ought to be competent judges, and perhaps are, maintain that it comes to the
same thing, no matter whether horses are tried at high or
low weights : and amongst those who entertained that
opinion was the late Lord George Bentinck, I believe.
Still, with all due respect, my vote would be against any
such idea ; for many reasons.   In the first place, all horses
are not equally formed, nor is their action suited to carry
heavy weights, and common sense must dictate the
absurdity of supposing that the higher they were set the
more they should not tell, especially when they exceeded
the acknowledged racing weight, 8 st. 7 lb.*   Who, if he

* Upon the subject of weights, an opinion which I entertain
may call up the ridicule of many, as regards the error of putting
up such *very light* children to ride horses in their *general exercise*,
in comparison with the weights they may be bound to carry in
their engagements; namely, that, as in most other cases, " practice
makes perfect, and habit becomes second nature," and, to a great

were possessed of the ordinary share of that sagacity with which racing-men are accredited, having even made a match, and given away a stone, would set the weights, at racing weight, if he had the option of doing so, at 7 st.? Those who entertain the opposite opinions should bear in mind the old adage, "the last feather breaks the camel's back," and remember the fable of

> " The donkey, whose talents for burdens was wondrous,
> So much that you 'd swear he rejoiced in a load,
> One day had to jog under panniers so pond'rous,
> That down the poor donkey fell smack on the road."

Those opposed to handicapping maintain that it tends to injure the breed of horses in general, which, if it were the only grounds of objection, would be a slender one indeed. It is not likely that breeders would take pains to produce bad horses for the purpose of getting them lightly weighted, nor is it the fact that they are weighted according to *size*; many of the *finest animals* are the *lightest weighted*, and if the adjudication thereof were taken according to the judgment of some of those who declare against the principle, their efforts would be attended with success so far only as to cause merriment and plenty of "fun" to the spectators. Men purchase young horses with high expectations, and if they had not something to turn round upon, in the shape of handicaps, in the event of their higher hopes being frustrated, they would not be so

extent, inures the frame to exertion: the fact of the animal being accustomed to a feather-weight, and suddenly loaded on the racecourse to contend under heavy weights, must, to a certain degree, have an injurious effect, independently of other objections, such as not being properly held together, and boring on the bridle.

ready to purchase ; and then the breeder would naturally
be discouraged, and down would go the value of the horse
in every respect.   But it seems strange, amid the great
outcry against the system of handicapping, that, amongst
the ranks of its enemies, there cannot be found any
bright enough to provide a substitute without diminishing
the amount of either sport or speculation.   The originator
of the one system may not have been possessed of more
intellect than others ; and it would be a great libel
on that of so experienced and so enlightened a body
as the rulers of the turf, if amongst them there did
not exist one capable of improving upon a plan which
appears to some unpopular, and attended with so many
drawbacks.

It has been frequently asked, "What is there in a
name ?"  There is a great deal.   The name of " *Selling
Stakes* " is not a favourite with either side.   Then suppose
the stakes were called 'The Prince of Wales,' 'The
Grand Alexandra Prize,' 'The Garibaldi Goblet,' or 'Tom
Thumb's Thimble,' and that the conditions were the
same, and the *secondary consideration* (?)—in fact, every
part of the terms, the same as they are at present in
handicaps, with the exception of the mode of adjusting
the weights; that the latter were put upon their horses
by the *respective owners*, in a *sealed entry*, and deposited,
the day of entry, with the proper authority, each owner
entering his horse, affixing to his name the price he was
to be sold for, whether for 3000, 1000, or 100, and at which
price he might be claimed according to rules framed,
or to be framed, by the Jockey Club; what difficulty
would there be, beyond the regulation of the sliding
scales of weights, in accordance with the prices, which

those authorities, it is to be presumed, would find no
more difficulty in arriving at than in the adjustments of
weights for Queen's plates? It may be said that there
would not be so many entries. That is very questionable: but if reduced they could not be by so
great a number as in a proportionate view would reduce
fields (if the stakes were worth winning) to so great
an extent as the non-contents usually found in an
entry of one hundred horses, or the usual number
entered for large handicaps would do, because those
who *would enter* would do so with a *knowledge* of *their
weights*. Such conditions could be framed as would
prevent any possibility of the objections to the former
conditions of selling stakes; and if there were no handicaps, owners wavering about entering under such conditions would be obliged to do so, having no other mode
of winning; while they would have the same chances as
ever of winning weight for age races.

The only difficulty would be as to the sliding scale.
Assuming that an owner would not feel disposed to sell
his horse *at all*, he must be a very valuable one, and there
are plenty of richer stakes for such animals; and, moreover, it is the very fact of such "swans" being entered
for handicaps that causes all the outcry on the part of the
owners of moderate horses, as well as their own, who are
always the most likely to complain if they are treated
according to their merits. The very entry of such
horses in such races, which, in fact, is not their place,
causes all the trouble and complaints of very light
weights: in short, they are the cause of the difficulty.
Some owners, perhaps after having paid 1000 guineas
for a yearling, find they cannot win *at all*—as there are

many horses at present could not win if let loose in good
company—yet they have no opportunity of getting
back any of their losses; whereas, if there were certain
*classes* formed for such horses, or for moderate ones, there
would be plenty of racing and plenty of profit, and an
opportunity for owners to get rid of such animals should
they feel disposed to fly at higher game. Amusement
does not *altogether* depend upon the *class* of horse;
nor does the amount of speculation either, although no
doubt there is most interest in witnessing contests
between first-class or celebrated horses. The authorities would not find it very difficult to frame scales
which would be attended with most beneficial results, and
be a very great improvement upon the present system.
If, for instance, any of those large handicaps were
called by one of those grand, sounding titles, and were
framed according to such terms, what would the odds be
that there would not be more entered, or *start* at least,
than appear at the post now-a-days?

The *true test* of the *value* and *merits* of *any* horse, as a
general rule, is the price his owner *will* part with him for;
and it is a much fairer one than that which may be *fancied* by any other party unacquainted with his merits, of
which, it is to be presumed, his owner is fully aware, at
least more so than others, although even in this respect
there are some exceptions. An eccentric friend of mine
used to be driven to a state of excitement, almost bordering upon distraction, at the very mention of the fact of
an owner knowing anything at all about his own horse; so
much so, that it was a regular joke amongst his sporting
friends to refer to the subject. What reasons he could
have had for arriving at his conclusions were, it is to be

presumed, best known to himself; he, however, seemed to have formed very fixed notions on the point.*

But *redeo ad rem.* As to purchasing a stud, or adding to one already formed, perhaps one of the greatest mistakes frequently made is that of buying *too many,* especially of a *moderate* class, for very many reasons: amongst others, the object may be to become possessed of *first-class horses;* and few possess the moral courage to get rid of the inferior ones, or adopt the maxim "that the first loss is the best." It stands to reason, that to get those rare acquisitions, *first-class racehorses,* the best mode is to purchase the most promising; and it is but natural to assume, that the greater number any one purchases, the more likely he is to hit upon good ones: but the great drawback is, that he thus encumbers himself with too many. Any owner anxious to accomplish his undertaking, should make it a rule to get rid of the others, when found wanting, even at a sacrifice, and if he could not do so otherwise, it would be a far more judicious course, under any circumstances, to give them away, than adopt that so frequently pursued, of holding on in hope of a handicap. Many are carried on from time to time, each reduction in weight rendering it in his mind nearer to the "*moral,*" as it is termed ; but that "moral" frequently turns out a "myth," ending in a break-down, or some other disappointment.

* The best bet I ever won was when I took, amongst others, one thousand to fifteen pounds about an animal the moment I saw the weights, and the horse started at a hundred to one, the owner at the same time backing another of his own, at very short odds : the former won in a canter. One of the greatest examples I ever knew in racing, of either the brightest display of generalship, or most palpable stupidity, ever manifested.

Any owner keeping a number of horses, if speculation be his object, can win as much money with a *few* horses and upon a *few* races as he can with a dozen. Some people, who really understand the prudent course in the various branches connected with the animal, and the management, placing, and other necessary matters relating thereto, would win far more with a small stud than others with a much larger one.* Some keep on day after day, entering, engaging, travelling, and perpetually backing them, frequently " *merely because they are their own;* " not that they think they can " *absolutely* win," but " they don't like to let them run without a ' pony,' " and so on ; or they " *only* backed them (when beaten) for a hundred, sometimes called a ' century:' " but this " century system," which sometimes is continued for a quarter of a century, if the estates have not been entailed, in any case amounts to more money than at first imagined, and all the time the heavy drag is kept on, the bad ones eating as much, and generally costing more, than the good. This may be best termed the " dribbling system," which never pays. The continual drag is too heavy. He who keeps a lesser number, and backs them at the proper time for a sum sufficiently remunerative, and otherwise adopts a course of prudence, which judgment and experience alone can dictate, is the party most likely to succeed. Judgment and experience will always beat money *in the long run;* for however the plunging prin-

---

* It would almost require Mr. Judge Clark's *fiat*, at least once in every three races, otherwise the decease of some dearly-beloved relative annually, to keep accounts square with the number of horses kept by some enterprising aspirants, assuming that fortune should smile occasionally on others.

ciple in any branch may prosper for a time, it never lasts. Sometimes horses are sold, from the fact that the owner has too many; and amongst them perhaps some good, which are purchased at " par " by some really sound judge : and this course is frequently caused by persons not being in a position to " see it out " with their horses. Where is the man who ever rushed recklessly into a *large* stud, and ever found it pay ? How many men of judgment, who adopted the other system, have made fortunes proportionate to their attempts or their aspirations ? And why ? because they, with comparatively little expense, have had plenty to work against, furnished by more extravagant and reckless owners. Have not even men who could be named (now *under* the turf) kept something bordering on one hundred horses at a time; and although really men of judgment, and with experience of the animal, yet made the " grand mistake" referred to ? It is wonderful that people do not more frequently ask themselves, " where is the money to come from to enable them, without losing their own, to even have their enjoyment ?" Of course it is a different case with those who can afford to keep a thousand horses if they choose, to *amuse* themselves, like the Sultan with his Harem, and the public ought to be very much obliged to them ; and most unquestionably the success of such owners should, and would be, at all times, hailed with delight by all true sportsmen, for, in every sense of the word, they are deserving of it. Still, in the case of others, who had an eye to something beyond amusement, the question becomes of more importance as to their probable success.

The El Dorado, in search of which most aspirants cast their bark, consists of certain value. The question is, what that value is, and the expense which the "fitting-out" costs, the various appendages thereto, and the number of "sailors" on board? An inexperienced mariner, unacquainted with the shoals and quicksands, can hardly be too cautious, especially of sharks, which frequently abound, and with their satellites, or pilot-fish, generally look after the best-conditioned prey, lurking nearest the spot where it is most likely to be found; and once they get the chance, escape is a miracle. The instinct of those fish, especially combined with that of their aides-de-camp, is surprising, their bite frequently fatal.

Probably the most important subject is as to the real value of the sought-for prize, and the comparative expense attending the realisation of it. It has always been admitted that the turf was remarkable for its uncertainties, its ups and downs, occasional successes to some, yet more frequent reverses to most people. It might, however, not prove a very difficult task to arrive at the principal cause.

The statistics of horseracing furnish materials, clearly demonstrating the fallacy of supposing that it is possible for it to have any other result than one of certain loss to the majority of its patrons, forming or framing calculations *merely confined to the value of prizes,* which may be won *independently of betting speculations.* Those racing records are framed and rendered in a style not excelled, if equalled, by any other public returns in simplicity and correctness. By reference thereto it would appear, that there are at present, and have been for the last few years, taking one with the other (independently of numerous others purchased, trained, &c. and found useless), from 1800 to 2000 horses, contending for a certain number of stakes or prizes, numbering about 1500. In calculating the pro-

bable amount of expenses necessarily attending the keeping of racehorses, and making a rough guess, it would
hardly be exceeding a fair estimate to rate each horse,
including stakes, forfeit, training, travelling, and the
numerous *et cæteras,* at 300*l.* per annum each (which
might in many cases be more properly set at 500*l.*).
Would a contractor accept of one million per annum,
and undertake to supply the deficiency to be found by
owners of horses annually? In other words, does that
amount cover the outlay of the owners of all the animals
that contend for races, as well as those proved worthless? What, then, would be the amazement of some
of the patrons of the turf if they were asked to take
a few shares in one of those numerous monster schemes
so frequently, in the present enterprising age, submitted for public patronage, if they set forth upon their
prospectus such enticing proofs of their desirability and
probability of success or remuneration? (The Earth-
quake-and-Balloon scheme will probably be the next
project.) Such statistics as these might be fairly urged
as evidence of the truth and justice of the remark, that
the "glorious pastime" was originally intended for, and
is best suited to, the nobility and aristocracy, who are
presumed to be not only the most noble representatives
of mankind, but likewise the possessors of the superfluity
of that much-prized metal, which they sometimes appear
to hold in quite as high estimation as their more humble
fellow-creatures. Then, if success depends upon betting
transactions, the question arises as to how far the absolute
*ownership* of horses benefits the *proprietor* beyond the
*public,* and how such benefits can be derived, curtailed,
or completely prevented? Assuming that an owner is

desirous of blending profit with pleasure, the very first step necessary to be taken is to secure the services of a *trustworthy* and *competent trainer*, with a *silent tongue* (not with a clapper like Big Ben); inasmuch as, next to becoming possessed of good horses, the subject most requiring attention, and one of most vital importance, is to provide himself with a good trainer. Still, a very far-fetched and exaggerated idea is sometimes entertained with regard to the superiority of one above another, so frequently remarked upon, even among themselves ; viz. that some are so far preferable to others, and excel to so great an extent in skill.

If one were reminded that it required great study to arrive at a fellowship or scholarship at college, the fact would be admitted as a matter of course, but to argue that there exists that wonderful art or science in training a racehorse is simply a farce, which may be rendered theatrical to a great extent, by the acting of persons whose extraordinary imaginative genius, and sometimes consequential and bland style, above that of their more unassuming brethren, frequently carry away the credulous to believe that it requires a wizard to win a Derby or bring condition to perfection ; almost persuading many to a belief that some trainers could win a Derby with a bad horse, like Baron Munchausen's harp, that "played the tune without the performer." Good horses make good trainers — plenty of good hay and corn tend wonderfully towards condition. The usual duties of a stable are nothing beyond that which the most humble individual may become acquainted with in time. The best "head-lads" make the best trainers, many being for years, in some instances, in reality the trainers.

There was a day, when the "good old fashions" would at a glance impress one with the idea of the man who really understood his business, and was not above it. Fortunately, many of the same class still remain. Some have been lucky in having had good horses placed under their charge, while others have been the reverse; the success of those animals tending in the eyes of some persons to stamp their trainers as "nonpareils," the consequence being an immediate run to fill their stables; while others, equally scientific, and perhaps otherwise more deserving, have been just as unfortunate in not getting the chance, as their employers have been in not obtaining good animals. That there is a certain knowledge necessary to be obtained, and which must be improved by practice, there can be no question; that it requires a sort of apprenticeship is true, the more humbly served the more likely to arrive at perfection. A certain form of condition must be arrived at to have the horse really "fit," still it does not require that wonderful talent which many people appear to fancy. It is true that, having learned the regular *routine* to be pursued in order to arrive at perfect condition, the more intellect coupled with practical experience any man may have been gifted with, the more likely he is to prove successful: but there is but one pitch of condition, "perfection;" and that is as often *overdone* by the "artists" as it is *underdone* by the "non-professors," or more humble members of the profession. In many instances the animals themselves, by nature and a mixture of chance, happen to be *exactly the thing* on the day, although having, perhaps, been a shade removed therefrom within a very short period preceding, for it is quite true that a sweat or a couple of good gallops will effect a wonderful change. Horses cannot, like pedestrians or prize-fighters, tell their trainers, or friends, "they never felt better, and are fit to contend for a kingdom." There are many instances where horses have "raced themselves" into condition previous to an event, their success for which has astonished none more than their owners or trainers, owing to their having been perhaps defeated previously, when their success has been looked upon as a foregone conclusion. The fact is, there is a deal of chance in condition as well as in every other matter connected with horseracing; for no trainer, be he ever so talented or experienced, can at all times have his horse as fit, or as well as he might wish. These remarks are made merely to impress upon the reader that there is more fuss made about the *science* of training than it really calls for; for admitting that there is a great deal of practical experience and intellect necessary, and as much difference in the appearance and condition of some horses as there is between those cherished relics of aristocratic antiquity at Newmarket and the public stand-houses at gorgeous Goodwood and elsewhere, still the opinions of required skill are frequently exaggerated.

In some instances trainers have a greater number of animals under their charge than they could by possibility properly attend to, if it were absolutely necessary that each should have their special attention. No doubt they have head-men or assistants, some of whom occasionally know as much, if not more, than the masters, and in reality are the trainers; and why should they not? In some such cases, no doubt, the best performers come in for the lion's share of *the* trainer's attentions, while the inferior are comparatively disregarded.

Trainers, as a body, are respectable, trustworthy, and intelligent, and many of them most independent. Although intelligence and experience are necessary, it by no means follows that an *illiterate* man cannot be a good trainer. Some of the best have hardly known how to write their names, whilst others have learned to write too well to be either an acquisition or beneficial to their employers. "A little learning is a dangerous thing" at times. Some years ago, a certain trainer whose horses had won Cæsarewitches, Liverpool cups, &c., upon being asked suddenly by his employer to read the weights of a handicap just published, and give his opinion thereupon, quietly handing back the paper remarked,— " There ! please to read them to me. I have been trying that game on long enough ; *divil* a word myself can read. The old woman (his wife) always *does* the reading and writing for me." On a previous occasion he had been found with a newspaper turned upside down, but he had discussed the weights previously with others.

A horse, trained by our hero, had won easily a certain celebrated race in England, subsequent to which he was immediately purchased, for a large price, by a patron of the turf, who called over the trainer, a very humble, although clever and experienced man ; the latter, snatching his *dudeen* from his mouth, and touching his *caubeen*, was requested by the purchaser to inform his trainer, Mr. So-and-So, as to the horse's constitution, habits, &c., adding, " that he believed, when he had received the 'polish,' he would ' do a good thing.'" " Oh !" replied the late trainer (who was not over-*polished* himself), " why should the *likes* of me attempt to give Mr. So-and-So any instructions ?" The fact was, the horse was in superb form at the time—a perfect picture of condition, admired by all judges, but having been removed to his destination was tried over and over again, galloped almost to death, and never won a farthing afterwards.

There are very many competent men to be found, whose very appearance would denote a fitness for their duty adhering thereto, and not outstripping it, who will not try your horses without your knowledge or authority; or try them at one time before your face with one incorrect result, again behind your back with a different one, leaving you in total ignorance of the result of the genuine trial, and turning it to their own advantage, and that of necessary and obliging friends, *sub rosâ*, who possibly, having made a few temporary advances, consider they are *entitled* to know the merits of an animal before even the owner himself. It is all very well to be possessed of good horses; the difficulty of finding them is best known to those who have tried the experiment: the disappointment which they so repeatedly experience, and so dearly pay for, all tend to test the *" staying powers"* of *owners,* as well as horses. But when fortune may have favoured an owner, be his position high or humble, with animals so rarely found— *" racehorses"*—the matter becomes somewhat serious when he finds all his hopes blighted, through the instrumentality, or connivance, of the very party in whom he may have placed implicit confidence. Let any man place himself in the position of an owner, who, putting his own interests aside, in every sincerity may have expressed to his friends his sanguine expectations of success; then, what would the individual deserve who, with perhaps an outward appearance of honesty and straightforwardness, " a lip of lies, a face formed to conceal," could act the

E

part of the frozen serpent, and turn upon his employer and benefactor, by a dishonest sacrifice of his interests?

> "And in my mind there is no traitor like
> Him whose domestic treason plants the poniard
> Within the breast which trusted to his truth."

On the other hand, the trustworthy trainer is invaluable to an owner; and to one who could afford it, amount of remuneration should be a secondary consideration, as he best studies his own interests by making those of his trainer identical with them; and next to an owner, the trainer's interests should be consulted and considered. Any proprietor of a number of horses would, for many reasons, adopt by far the most prudent course, by having his private trainer, or, at least, one to act conjointly with a friend or two; for one of the great objects to be attained is privacy in such matters, as well as quietness for the animals. The frequent visits of the various employers (some, perhaps, with a fourth share of a leather plater), with their hosts of friends, one after the other, morning, noon, and night, stalking about the stables, puffing their cigars, worrying the horses (sometimes kept stripped and tantalised by rubbing an extra half-hour, during the pleasure of their admirers), taking their notes, and circulating the number of coughs; "pumping" attendants and little boys, on various subjects; in short, "poking their noses" where they are not required, and should not be admitted, together with the fact that the trainer is kept in a perpetual state of bewilderment, answering either absurd questions or evading others with *more meaning;* and taking into consideration the fact that a man with a lot of horses to attend to properly has quite enough on his mind (if not too much), and requires a little rest; all must tend to render matters anything but desirable or beneficial. In fact, the doctrine is erroneous; and any owner of horses who can afford to keep a private trainer, or join a friend, with whose interests his own may become identical, makes a very great mistake indeed in adopting the opposite course; which he will only find out when he has been worried to death in many ways, and, though last not least, obliged to back his horses at about one-half the proper odds: for there are sometimes a drove of followers in stables, who, from a sort of custom, appear to fancy they are quite as well entitled as the owner to know every movement (and sometimes know more), and be " on," at the best odds. Thus owners are frequently driven to " scratch" their horses out of their engagements, simply from the fact that they have been forestalled by people who make it their business to become acquainted with facts, which they communicate to their respective connexions: the result is, the owner in disgust strikes his horse out, because he could not get his own money on, and then comes the thunder of abuse from those who, enraged because horses were not kept for their use and benefit, express themselves in terms anything but complimentary.

It is all very well for people to rave about proprietors who keep horses for " honour and glory, and that sort of thing;" there are a great many more who " wait for the waggon with the universal idol," and like a little of the " cream" of the market, instead of being obliged to take the " skimmed milk." Still it sometimes happens, that the very people most closely connected with stables, and who *ought* to be the *very first* to know when there existed a probability of success, are not only the *very last* to learn it, but absolutely made the useful instruments, in the shape of cats' paws or jackals, in order to carry out "arrangements" suitable to certain parties, who may have been accommodating, under peculiar circumstances. Such proceedings are not, happily, of frequent occurrence, and are not practised in respectable establishments, where the interests of owners are really studied. It is also very well to run away with ideas, as to the advantages one has in being the owner; but what are they? and where do they begin and terminate? Those benefits or advantages may be great, or may be quite the contrary. He has to find the animals, and all the expenses attending them; and the entire benefit consists in the fact of his having not only *more* knowledge of the merits of his horses, but a *prior* one to the *public*.

Suppose an owner has purchased a splendid stud, upon a scale of magnitude; or, on the other hand, take the case of what is commonly termed a "little man," whose principal, if not entire dependence, rests upon his success. Then to what possible end can all this tend, if the cup of success is to be snatched from him through the ignorance, ingratitude, or dishonesty of the trainer? It therefore becomes absolutely necessary, and of the most vital importance, to find a man to whom the responsibilities and care of such can be entrusted, and in whom implicit confidence can be reposed; as it is like placing one's capital in a safe, and handing the key to another. Therefore, in order to avoid the possibility of one being like Tristram Fickle, in the farce of "The Weathercock," changing with every breath of wind, which frequently ends in a storm of suspicions and want of confidence, it is absolutely necessary to secure the services of a trust-

worthy trainer; without which he might as well carry
one of "Dent's best" without a key to wind it, and
all the 'Flying Dutchmen' or 'West Australians' that
ever were foaled would not only be useless, but ruinous.
When confidence dies and candour departs between employer and trainer, rather than continue such a course it
would be more prudent to sell for whatever they brought;
and if purchasers could not be found, open the stable-
door, turn the animals loose, and let the first who caught
them keep them; unless the owner wished to live to find
the hair of his head make its appearance through the
crown of his hat, his ancestors having neglected to entail
the estates.

The steady, unassuming, and industrious man, is the
one to select: examples are to be found frequently.

" Let them talk as they will about writing and reading,
  And science in training, the chief thing is feeding;
  Strictly trustworthy, a son of old Dumbery
  Is the trainer—believe me, the rest is mere flummery."

One of the most important necessaries, in addition to
a properly regulated establishment, and a principal link
in the chain of success, is a good and true " trial horse;"
one which, as an old trainer of mine, dead many years ago,
used to say, " would tell one to a second what o'clock
it was." The want of this most important requisite
is invariably the cause of owners losing large sums;
and moreover, even when such necessary tell-tales are
supplied, it is not every owner or trainer knows how
to use them. On the contrary, there are more mistakes
made in trials than many people have any idea of; and
to my mind, it is not only one of the chief parts of the

science which trainers should be versed in, but it is the very one which they are most frequently, as well as their employers, least conversant with. Some are more skilled and preferable to others in this respect, as well as in judgment and knowledge of the animal, than in the mere matter of condition. Such mistakes are made most frequently through mismanagement as to jockeys and pace, as well as from the weights under which horses are tried, frequently differing so much from those under which they may be about to contend in their engagements. In many instances experienced jockeys are put on some, mere lads on others. Again, the pace half the way is frequently little more than half speed; the boys sometimes chatting to each other. Then some horses, especially old ones, become so accustomed to particular ground, that they frequently do not really try; or if they win, they *merely* do so, leading to the supposition that they *only just won;* and orders are frequently given "not to abuse the old horse." An instance of this kind occurred, and which might have been attended with serious results, and, as it was, attended with one anything but gratifying, beyond a temporary delusion that a small mine of wealth had been discovered, in the shape of half-a-dozen two-year-old flyers, which was soon dispelled. Having tried several times a number of two-years-old with a well-known old one, and some others, at certain weights, the result on each occasion was exactly the same—the old one *just winning* by a length, the rest in a "lump." All wonders, of course! although, in the latter respect, the matter did not look well. Thanks to the intelligence of my respected and experienced trainer, Mr. H. May, a change of ground being decided upon, from a belief that there were too many *flyers* in the *covey* to be true, we tried the reverse way, running in the direction of the " corn-bin;" when the young birds' wings were woefully clipped, and they were left scattered in all directions, some not within gun-shot of others, for the old one might almost have had her share of the oats consumed before they arrived for theirs.

The ground has a great deal to do with the correctness of a trial; for some horses like certain gallops and dread others, from having too frequently had the " persuaders" applied. The ground, if possible, should resemble in every respect that over which the horse about to be tried is to contend for his engagements in public; and in every respect, in fact, the *private trial* should resemble the *public race* — jackets, tight light saddles (which frequently frighten young horses, if unaccustomed to them), &c. *The more remote from their usual exercise-ground the better.*

Some horses will almost fly in private, either with or without their clothes, yet when they are stripped in public will literally die under the jockey, or, upon the appearance of a crowd, never try at all.

An extraordinary instance of this occurred some years ago with a horse of mine, which probably no animal living could have beaten in private. This horse (he was a gelding by ' Harkaway,' and half-bred) led both myself and trainer to fancy, either that the others, although they had previously won and defeated easily all the horses of their year in Ireland, were moderate, or that he was a perfect wonder. Having entered him for a few handicaps, and upon the first occasion backed him, as if the result were over and decided, he was not only defeated easily, but was absolutely last — beaten a hundred yards, to our great astonishment and the

amusement of others, who enjoyed the thing amazingly.
The other horses with which he had been tried ran
subsequently during the day, and, not backed for a penny,
won as they liked. The following day, having resolved to
give him one more chance, and being laughed at by some
friends who were rather disappointed at having lost their
money, the brute, with extreme odds against him, absolutely won as far as he was defeated the day previously,
beating a large field and upsetting a "tremendous pot"
backed against the field. The fact was, just previous to and
during the race a heavy shower of hail came down; everybody took shelter therefrom; the course appeared quite deserted, and in came the big seventeen-hands-high "buffalo"
by himself (he had a pair of horns about two inches long
on his forehead). The result led to the "usual remarks"
upon all sides; the most wounded sufferers of course being
the layers, and the *friends* who were not *on* at the proper
time, according to their ideas and wishes. But such are
frequent cases, and prove the necessity as well as value of
a genuine trial horse. This animal turned out a first-class
steeple-chase horse, although a coward on the flat. The
true worth and staying qualities of horses, like those of
their masters and friends, are best tested and tried when
in difficulties; the mistakes so frequently made, and
delusions under which so many labour, might in many
cases be obviated, and the results be rendered more harmless, by an observance of that caution so often disregarded.

But the prudence of trying horses *at all*, at least to
that extent so frequently practised in cases where a great
event or large stake is in question, is one admitting of
doubt, where the object is not to win largely by the result;
for in many instances the consequences are most detri-

mental, and sometimes fatal, extinguishing not only the *immediate* chance, but putting the animals completely *hors de combat.*

Suppose an owner has already backed his horse for a large stake, say for a Derby. On the eve thereof the animal is tried, and asked a Derby question, which he finds very difficult to answer. Although favourites invariably do so, as far as report—people seldom hear of one *losing* his trial. Then in what better position is he, further than that he has the pleasing gratification of a further belief that he will find his name recorded as the owner of a winner of the "Blue Riband?" If the result turn out an unfavourable one, the wires are certain to be at work from some quarter or other; and the metallic influences are also, in other respects, in requisition and full play. Some very experienced patrons of the turf completely set their faces against the system, merely satisfied with a "rough-up." One of the best judges and most experienced and successful men, in every respect, in breeding and racing, did not approve of *absolute trials.* He was not a heavy better, generally confining his investments to a ten or a twenty-pound note. He had an observatory specially constructed, from which he, with the assistance of a telescope, witnessed the daily exercise of his horses; and upon the eve of a race-meeting had them sent from a certain point, a given distance, in their regular Indian file, finishing opposite his post. If asked what he thought of his chance, he would invariably reply " that his horse ran untried :" but whenever his *ten* was on he generally won; and if a "*pony,*" it was all over. He was, in my humble opinion, the best judge of a horse, in every respect, and everything relating to breeding and horseracing, that ever lived — (a large word, no doubt) — and many an hour's amusement and instruction he afforded those with whom he was acquainted, having been, in addition to his other qualifications, a V. S. of nearly half-a-century's first-rate practice.*

One of the most important matters to have regard to in the management of horses, and one upon which success principally depends, is the *placing* and *engaging* of them. It is much better to be *first* in *moderate* than *second* in *superior* company; and in this the judgment is best tested. As Cæsar remarked, "that he would rather be *first* in a *hamlet* than have an *equal,* or be *second* in Rome." The wholesale manner in which horses are sometimes *crammed* into engagements, especially when the option of doing so is entrusted to parties who have not to pay the forfeits, is something surprising. In many instances, the absence of owners' names from a sheet calendar or race-card is as rare an occurrence as the appearance of a woodcock in summer, or a policeman when really required. How horses, especially young ones, can be expected to fulfil those numerous undertakings, is a matter best known to those who adopt the system. Prejudice of some owners in favour of their own, and belief in their powers to vanquish all comers, being the very essence and life of horseracing, and when carried to excess, which it so frequently is, the stumbling-block to success. Some

---

* All horses will not gallop in their clothes; still, a very fair estimate of their qualities may occasionally be made under such circumstances: a contradiction to which opinion would be a serious libel upon the talents, as well as detrimental to the pecuniary interests, of a certain professional fraternity, forming a large body of the *cognoscenti.*

owners (especially beginners) fancy their "geese" are all "swans," invariably assuring their friends on the eve of an event that it is a certainty for one of their stud, whether for 'The Devil among the Tailors,' 'All Round my Hat,' 'Tickle my Fancy,' or 'Impetuous Bess;' an attempt to change them from which infatuation would prove as futile as to persuade a young lady to dispense with her crinoline. It would, however, be prudent to bear in mind, that others likewise have good horses, as also the dangerous results arising from being over-sanguine. Disappointment to such parties, who are generally of a rather excitable temperament, sometimes leading into the river of revelry, which not unfrequently flows into the sea of adversity.

The plains of Newmarket afford to the beginner ample opportunity of displaying his taste, gratifying his ambitions, as well as developing his resources in his favourite pastime; still, although it may be the admitted "metropolis of horseracing," yet it would hardly be the spot likely to be selected to make a favourable impression on those wavering in their opinions as to the superiority of the turf over all other pastimes : nor does it unfold the most pleasing representations in many respects.

Goodwood ! — Glorious Goodwood, Beauteous Brighton, Royal Ascot, or Metropolitan Epsom, would be more likely to have the desired effect. Goodwood, from its elevated, yet sheltered position, unfolding to the wandering eye, amid her hills and valleys, the richness and beauty of her woodland scenery, rivalling therewith her sister, Beauteous Brighton, the queen of watering-places, with her varied views of land and sea, the " blue above and the blue below," her bracing breezes and marine retreats ; they from their proximity vie with each other in excellence, but each aid in restoring to the mind and body those invigorating and soothing influences so much needed, and so frequently impaired, especially by the zealous patrons of the "glorious pastime."

Ascot, patronised by Royalty, where the eye of the visitor will at once become fixed in admiration upon the array of beauty which so adds to the scene, by the presence of the fair sex — those charming objects, eclipsing the other beauties of Nature,—

> "In lines of light, beneath the golden sunbeam's hues,
>     Like stars through heaven's sea  ·
>   Floating in harmony,
>   And casting a lustre of light to all around."

All combining to render the contrast with the "Metropolis" most striking, and with anything but a tendency towards raising the latter, even in the eyes of its most zealous admirers.

If, indeed, the opponents of the turf sought for a picture whereupon to paint, if not its defects, at least the absence of all additional charms beyond mere speculatory recommendations, they could hardly select a more favourable one than Newmarket.

It is quite true, it at once unfolds in its outward and visible *tout ensemble* the inward and real meaning of its objects, without affording beyond the mere arena for the development of its purposes any additional prepossessing recommendations. As the Head-quarters of horseracing, and as the Court wherein its laws are framed and administered from, it stands "alone in its glory;" but even in the benefits or advantages necessary to the advancement of its objects, such as its training-grounds, it can hardly boast of equality with, and much less with superiority over many others. It cannot be denied that, in its *abundance* of racing, it excels all other meetings; although it neither surpasses nor equals many others in the *superiority* of sport. In its *superfluity*, it is almost sufficient to remind one of the old, although not very refined adage, that "too much pudding would choke even a dog." The solemnity, as well as precision (rivalling Costa, or Jullien in his best day), with which some of its rules are carried out, are remarkably business-like; perhaps, in the extreme, almost sufficient to impress — suppose a foreigner, who might not be conversant with the English language or habits—with the idea that he had come not to a *rendezvous* of racing, but to a more solemn duty; for he would be rather surprised upon an evening (perhaps a Sunday), about nine o'clock, upon his *entrée* to an edifice resembling a Methodist meeting-house, to hear the weights of the various handicaps, as long almost as one of Blair's Sermons, read aloud, in a clear and clerical voice, to an attentive and anxious audience, without the soothing assistance of even a cigarette, so highly appreciated in continental countries.

The rush at the railway terminus upon the arrival of the afternoon "express," on a Sunday afternoon, bears a striking resemblance to the representation of Brown, Jones, and Robinson, looking after their baggage. Every man, who cannot afford the luxury of a valet, keeping a sharp look-out for "his own," which he so hastily conveys to his usual domicile, seldom remarkable for the moderation of its rent. Everything seems " money," racing times. What care racing-men what they pay? They *find* their money; the streets of Newmarket are paved with gold. Then, where is the race-card? generally as long as a lawyer's bill, printed in a *peculiar* kind of German-text—peculiar to the place, and looking like Greek to a stranger; yet Jemmy de Vergy can read it and mark the winners, with his " tip" into the bargain : he appears to monopolise the principal custom. The trade must be profitable and flourishing; for any person to witness the " rush" at the printing-office during any race meeting, would hardly require a better description of that at a railway terminus, at cockcrow in the morning, upon the occasion of an international prize-fight. The scene is one by no means the least amusing, in connexion with the sport. The crush of crinoline upon the occasion must be tremendous, if those fair retailers of the " correct cards" patronise that fashionable, yet deceptive, addition to the female figure.

But Newmarket, with its trifling drawbacks, must be always held in high respect, as the theatre at which so august, select, and zealous a body, as the " rulers of the turf" assemble, who are so chary of its interests and welfare, and without whose patronage and support the princely pastime could not be preserved in all its grandeur, but would unquestionably dwindle into insignificance. The air must be most beneficial to the health and soothing in its influences, for it is considered by those who ought to be best judges conducive towards enabling reporters for the press to carry out their arduous duties, even in the Cambridgeshire week, without further shelter than the celestial canopy. The patrons of sport are invariably fond of good living; and the casual visitor, if an epicure, should be made aware of the fact, that as Yorkshire is celebrated for its hams, Cheshire for its Dee salmon and cheese, so

Newmarket is renowned for its "pork sausages." Many a bottle of prime old port, Chambertin, Cliquot, &c., has been and will be uncorked in that little sporting village. The eventful anniversary which represents the discovery of this wonderful and anxiously-sought-for El Dorado, the prize so eagerly coveted by its numerous followers, and which casts such a halo of glory around the fortunate victor, is the signal for commotion amongst all classes, from the highest to the lowest in the land; each, to a certain extent, feeling a deep interest and anxiety, which no other event could create or bear comparison with. As the day approaches which is to decide the fate of the followers, and to whom the golden apple is to be awarded, everything is rife with excitement; everybody becomes more anxious to learn from his neighbour his opinion, whatever it may be worth, or however ignorant he may be as to the probable result; exchanges of which take place in the conference, more remarkable, however, for their extreme diversity than their value.

The chief fountain from whence the genuine "tip" is most likely to flow, when properly or legitimately "pumped," and into which trickle from various quarters the purest streams—in fact, a certainty, as far as human form can render it, and where, if it fails, it is looked upon as a sort of phenomenon in the history of the reign of "His Majesty"—is sought for in the neighbourhood of the "Green Park" (not always very green, even in May), where the highest in the land seek the proper path, and endeavour to ascertain in what direction his Majesty waves his magnetic wand, which sounds the tocsin, the signal for the rush of the multitude in the right direction, and generally seals the doom of many a desponding

aspirant. All who anxiously seek the favours of the
" King," in token of their loyalty, combine by volunteering their aid to increase the stability of his throne, by
each unfolding his stock of private information from his
knowledge-box; which is at all times most graciously
listened to (if not "taken in"), with that condescension
so truly characteristic of him who so generously and
truthfully, *at the proper time*, when consistent with the
duties of his important position, dispenses to his faithful
subjects the tickets for the " express," which invariably
wends its way, without danger to its passengers, in the
right direction.

> "The Lord and the Squire have a ' good thing' for Freddy —
>     A ' dead certainty' each for the next Derby Day ;
> But I'd wager a ducat he knows it already,
>     For his Lordship no doubt is a ' king' in his way."

The veteran sportsman calculates his age by the number
of Derbies he has witnessed, and as the Yorkshire squire
counts from ' Filho da Puta,' so others do from 'Whisker,'
each, on the eve of the coming events, discussing the merits
of many a flyer over their Chambertin or old port, and
the difference of opinion is thus happily settled. Many
friends meet on that occasion, who part only to renew it
under the same auspices.

Although the " Blue Riband" is looked upon, and is,
in fact, the greatest race as well as most valuable in amount,
and therefore the prize most coveted, and although the
winner is an animal of great merit, still it does not follow
that at all times he is the best horse of the year, nor yet
of those that absolutely start. On the contrary, comparatively speaking, and taking into consideration the great

weight attached to the victory, it is frequently won by moderate animals, and is a much easier task than generally supposed. For instance, it sometimes happens that many of the very horses that start, if not absolutely unfit, are suffering from the effects of recent attacks of influenza, or other diseases so prevalent during the long winter and trying spring, so fatal to horses in general, varying in severity as well as their nature. Some are seriously amiss on the day, yet started for various reasons; such as the "off chance," being superior horses, and the probability of others meeting with accidents. Others are debarred through a sort of "metallic fever," the most fatal of all; of which, when symptoms present themselves, it is most desirable to take precautionary measures, by becoming as far acquainted as possible with the operators who prescribe for them, who are generally most experienced practitioners, and keep them alive as long as possible, or human skill can avail. The patients seldom *absolutely die*, and often even *start* but with faint hopes of success, although some sanguine owners frequently are led to fancy, even to the last moment, that a spark of life remains, and nourish a hope like " the wounded soldier, struggling to the last."

The " winding-up" system, too, proves fatal in many instances, a break-down being a far more likely announcement on the eve of a Derby than the defeat of a favourite in a private trial; the frequent remark, " Such a horse will never stand a *Derby* preparation," being often verified.

It is a remarkable fact, that the Doncaster St. Leger winners have been far better horses, both as racehorses and at stud, as a lot; many winners of the latter having

defeated the Derby winners of their year, but yet
have been nowhere in the Derby. 'Stockwell's' year,
when 'Daniel O'Rourke' won the Derby, was a remarkable instance; for, without exaggeration, it would not be
going too far to say there were several far superior horses
behind him. In 'Blink Bonny's' year, in my opinion, the
best horse I ever saw gallop was not even entered for the
Derby or St. Leger, viz. 'Vedette.' What would he
have done over York, Doncaster, or Newmarket, with the
lot, for even the Derby distance? and if two miles, what
would the result have been? If four, he would have won
as far as a will-o'-the-wisp in an Irish bog—they would
perhaps get a glimpse of him. Without meaning for
a moment to deteriorate, or presume to lessen the qualities of that renowned, yet, to my mind, fortunate
mare, still I merely make the remark in giving expression to my humble opinion as to the animal, which
I consider the best horse within my memory, not
even excepting the great " Wonder of Erin," ' Faugh-a-
Ballagh,' with whom, if he were about to start for a match
over the Cæsarewitch course, my preference would be for
the former.

The name of winning the Derby goes far towards, not
only immortalising the owner (he should be created, if not
a Peer, at least a Knight; but certainly, if he " threw in
three mains," he should be raised to a seat in the Upper
House, considering that so many of those high personages
have so repeatedly failed to win even one), but has a very
great influence in many respects; amongst others, it tends
wonderfully to enhance his value as a sire, at least in
*prestige* if not in reality, having a powerful weight in public
opinion, although many a superior animal has been passed

over as a sire that would have been at the top of the tree had he *won the Derby;* while others, comparatively speaking wretches, have been patronised, and many of the best mares put to them : the consequence being the loss of the value of the latter by injury to their reputation. In any case the merits of the winners of the Derby can only be judged so far as that distance, for there are many animals defeated, especially over that course, that, if they had to contend for a longer or shorter distance over different ground, would defeat them. Every horse has a certain favourite distance, and a few lengths beyond it tells one way or other. The grand secret is to discover what that distance is, and innumerable are the mistakes made thereupon, and the hasty conclusions formed, totally at variance with their *real forte.*

But " the Derby day " having at length arrived, what a scene! A complete revolution of everything ! "The great city" becomes deserted, looking as if the long-expected French had reached Gravesend and the Thames been set on fire ; the Lords and Commons closed ; Rotten Row a "blank;" the World and his Wife gone to the Derby; the Jews supposed to have gone to Jericho. The Regent Street emporiums of silks and satins, pretty bonnets for prettier faces, blending and displaying the beauties of the rose and lily, and Fortnum, Mason, and Co. transplanted to Epsom Downs. Even the most needle-pointing snip would never entertain the absurd idea of sending his dun for " his little bill." Mr., Mrs., and the Misses Naggleton become a happy family, and join in the festival without *a word* of difference, except in the selection of a " favourite ;" and Master N., so sunk in admiration, and beginning to feel an inclination to become a candidate for future honours, if not a proprietor of a noble stud, asks his pa to buy a pony. The absentees are Methodist preachers, old maids and their lap-dogs, and the well-known firm of "Bobby and Cabby," none of the members of which would be in London if they were required. The scene presented on the Downs about two, when the great struggle is about to be decided, has but to be witnessed — description hardly requisite to any reader. Talk of Garibaldi in his red jacket, and his infatuated followers anxious to grasp the hand of the disinterested hero of so many fights (the last nine-days wonder, the forerunner of General Tom Thumb), even at the expense of the weight of a Bobby's bâton on the head, so mercilessly dealt out on the recent occasion of the King of Caprera's visit to the hospitable shores of Great Britain ! what comparison would their impetuosity bear towards that of the crowd, in their anxiety to get a peep at the probable winner — " the Crack," who not unfrequently carries triumphantly to victory the Garibaldian colours — the cherry jacket ?

"They 're off !" "They 're off !" "Hats off !" "Hats off !" There they go, and climb the hill like a herd of deer; 'Daniel,' 'Sunbeam,' 'Caller-Ou,' followed by 'Van Tromp' and 'Beadsman,' bang in front. What a string ! Now they near the bend ! 'The Dutchman' and 'Voltigeur' together down the hill, with a tail as long as the comet of '60. The pace has already told its tale. The pepper's out of 'Saucebox;' 'Daniel' gone to judgment (not by Mr. Clarke), his eagle wings already clipped. Tattenham Corner rounded; some have cracked. Hats off ! hats off ! Glasses up. The game is up with 'Gamester.' The 'Knight' has cast his die. The 'Merry Monarch' looks

*doleful,* and no longer goes *forth,* but yields his place to
'The Baron,' who gamely keeps his coronet and helps his
countryman to fight for the crown alongside 'The Marquis,' led by the gallant 'Caractacus.'  'Pyrrhus the First'
amongst the last condoling with 'Sir Tatton,' the Yorkshire
pet, whose happiest friend is the eccentric "Bill."  The
splendid 'Sunbeam' shines no more, but casts her rays of
hope on her game companion, the sterling *mine of wealth,*
the son of old 'Alice.'  They near the distance!  "Hats
off! hats off!" ("You, sir, in the white tile, get down off
the rails!" by some anxious speculator, but non-spec-
tator).  "What wins?  What wins?" (Another individual has just pulled down a fat gentleman from the
top of an artificial structure, and upset half-a-dozen.)
"Where's Faug?" (from the Emeralders.)  "Where's
the boy in yellow?  Where's 'The Dutchman?'  Tom,
with a grim death-hold of 'Ellington's' head, comes with
a rush; his backers gasping for breath, and turning all
the colours of the rainbow.  "*Honest Tom,*" cool but
sanguine, taps his box, takes a pinch, and halloos "Mine
wins!" Mat smiles.  'Thormanby' is there, alongside 'The
Dutchman,' upon whom Charley, his pilot, sits as steady
as a rock next the stand, anxiously looking out for his
friend Job on the game little chestnut, who gamely strug-
gles for the Cherry, and has "astonished his friends
the Browns."  'Beadsman's' beads seem nearly counted,
'Musjid' helping him; 'Van Tromp' right in front, using
his *broom* like a *brick:* there seems to be no end to it—
always at work.  Nat on the great big bay, clothed in
yellow surplice, begins to feel nervous; as does poor
old Isaac, whose cigar has almost become extinguished.
His lordship's confidence remains unshaken.  There's

young ' Clifden' by his side, looking as well as anything ;
but not a smile from ' Johnny.' ' Macaroni' looks like a
nailer about to hit the right 'un on the head ; his pilot
looking steady as usual, as if he had them dished. ' Im-
périeuse' is humbled ; ' Warlock' cries *" Peccavi !"* The
distance reached, ' Orlando's' done ; ' Blink Bonny,'
blinked, surrenders to ' Caller-Ou,' who cries " very fine
oysters, but no Queen's plates ;" resigns the office to the
bold ' Blair Athol,' to halloo the nine-day wonder. ' Kettledrum' on the lower side, bang in front, upholds the
fame of ' Old Ratty,' and rattles away to the air of
" Bonnie Dundee," and the fastest Derby on record.
The ' Voltigeurs ' and ' Cossacks' come with double-quick
pace and gamely fight their way, Alfred on ' Andover,'
with scientific hands and head, comes creeping up : but this
is not his *day.* The Wild Dayrellites discover that this
is not the Derby of '55. The struggle has commenced in
earnest. Shouts rend the skies from all sides. Here
comes Simmy the sensible, on ' Newminster,' sanguine,
and looking as if the *coals* were going to Newcastle ; the
days of ' Bay Middleton' flash across his mind : but so
does ' Stockwell' by him, who comes like a *thunderbolt*
from *St. Albans,* so near. Shouts from the aristocrats'
stand — " Exeter wins !" " No, he don't !" " Yes, he
does !" Bell bellows, "Clear the way !" but they won't;
he 's not the *perfect cure.* The Emeralders frantically shout
" Faugh-a-Ballagh !" throw their *hats up* in the air *for
joy,* and knock their *friends down* for *love.* The stand is
reached. ' Thormanby !' ' Thormanby !' ' Thormanby !'
from a thousand, as he comes stride by stride, looking as
if he'd like another round. ' West Australian'—the Australian, the " sombre gentleman in black "— comes with his

terrific rush. No: 'The Dutchman' wins. 'The Dutchman!' 'The Dutchman!' 'The Dutchman' unfurls his sails, and like a shot from the Kearsage snatches the prize, and lands the tartan and yellow to the dismay of the Australians. The "Blue Riband" goes over the border, to enliven the bonnets of Bonnie Dundee. Charley returns to scale with his simple yet unexulting smile, followed by poor Frank, looking graver than ever, who never will smile on earth again, and whose superior as a horseman we have never seen.

The race over, the various speculators of all classes discuss and inquire the effect the result may have had upon their respective financial departments, the following being about the probable replies to anxious inquiries from friends,—

*Blustering Bookmaker (half frantic).* " Won so much, can't guess!"

*Bumptious Bookmaker (hands in pockets, jingling sovereigns).* " A mere nothing; only won three thou——"

*Sensible and unassuming Bookmaker.* " Just got out, and a little the right side."

*Ravenous Backer (looking sad).* " A regular facer."

*Cautious Backer (who went to the Insurance Office).* " Five thou——"

*Clever Backer (besides his book).* "Enough of blunt to buy a brewery."

*Backer (who always wins).* " Just got on at the last moment."

*Unlucky Backer (who never wins).* " If I had backed him, he 'd have tumbled down."

Now for the champagne corks, which fly in all directions, the report resembling one of Garibaldi's hottest contests, generally issuing from ambuscades not only *manned* by many a devoted admirer or love-sick swain, but assisted by the objects of their most ardent love and admiration, who so bravely face and share in the danger consequent upon their return from the scene on that eventful day, amid a shower of grape in the shape of flower-bags, and shells of tiny representations of maternal care.

Then on to Cremorne, the next scene of battle so frequently selected, when the toils of the great day have compelled the Naggletons and Caudles to retire, and renew behind other scenes any difference left unsettled. On to the charge! Down go "The Blues!" the first Redan is taken, to the dismay of the bold and enterprising proprietor. Off go the few prisoners, the foremost in the fray, to hear the sentence of the tribunal; who, with the magnanimity of the true Briton, merely does his duty with a palliating remembrance of the Derby day.

All becomes serene; the spoils of battle divided among the victor and his friends, and his Majesty hailed upon all sides with acclamation, and assurances of the number of debts of gratitude due by his numerous friends. But some less fortunate, who have not belonged to his ministry, or had disregarded his advice, or joined other confederates, in mournful tones exclaim,—

> "Oh, my dear Freddy! how oft, if I *would*
> Have adhered to your counsel, I might have done good,"

are occasionally obliged to seek Solomon Sixty-per-cent; to whom they are introduced by his friends Flatcatcher and Touser, and received with the bland smile of welcome and twinkle of the eye, denoting an anxious readiness to accede to their requests, but regrets he cannot do so, having already promised an old customer, Captain Fireeater, who had called upon him the week preceding, in anticipation of the consequences of having backed three 'winter favourites'—two already dead 'uns;' the other, although not a 'corpse,' then in a very declining state. *Still, being desirous, if possible, to accommodate*, if they could induce their friends Lord Go-the-pace, or Sir Samuel Smashall (to whom, by way of parenthesis, he should very much like to be introduced), to affix his autograph to the "little instrument of security," he would *strain a point*, and *endeavour* to obtain *through a friend the few thousands required*, which would be accomplished at a *very great sacrifice.*" The job is done, and Solomon retails the bullion obtained at twenty at the moderate price, from his partiality to which he has derived his cognomen; and thus bids, for the time being, his new acquaintances, with their happy-to-see-you-sorry-to-meet-you countenances, *Bon soir—Au revoir !*

But there are others who have been deeply interested in the results of the great race, and have their rendezvous for settling their differences and discussing the merits of their respective favourites. Sam Shandigaff, of the sign of "The Racehorse," where all sporting events are on the *tapis* and ably analysed, the host (nicknamed "Mysterious Sam," from his policy of *taking in everything* and *letting out nothing*), has a large snuff-box (a gift from his grandfather, who trained for some of the "real old squires," in former days), with a painting on the lid, representing a human head, a mouth with a padlock, a pair of ears like overgrown mushrooms, and eyes like little Jack Horner in the pantomime, eating his Christmas pie. Sam, who is a

sober fellow himself, but likes to see his customers glorious
and excited over their differences as to their favourites,
merely sits in his arm-chair beside the fire, smokes his
long pipe, giving an occasional powerful whiff, resembling
smoke from the window of a house on fire, or a shunting
steam-engine, at the mention of any absurd opinion, sig-
nificant of the fact that he " knows a thing worth two of
that ; " merely makes use of the spittoon, and occasionally
amuses himself making paper pipe-lights.    There is old
Squeezem, the lawyer, in the corner, just dropped in,
looking as if he had lost a lawsuit; cursing " those pots
that always boil over.    He'll stand no more of Peter
Polish and his pots; he can't train a Derby winner—not
he ! he is only good for plating purposes.    His horses
sometimes *look* better outside than they feel inside : he
don't give enough of pepper.    What vexes him most is,
he put on Mrs. S.'s ' fiver,' and invited a dozen friends
to dinner; she having already chosen a splendid *moiré
antique,* from the sanguine hopes expressed by her friend
Mrs. P."

   " What a confounded fool he was not to follow Jack
Wilson's advice !    He always said ' this horse would do a
great thing some day ;' he was not fit in the ' guineas;'
and, as Jack said, " you can't have two bites of a cherry."
And he was right; he has won a ' regular stinger.'    He
said at the time, Sammy rode his head off in the ' two
thousand,' and that if I didn't back him he'd never speak
to me, or tell me a good thing again."    (I heard of a
" tout " once telling a nobleman the same thing.)

   But in comes Tommy Tightfit, the tailor; he has won
a " reeker," (he builds for the boys).    Sammy Sharpspurs
came to town last week and gave him the right " *tip.*"

" He rode in the trial; never crossed such a ' *tit :*' he rode the young 'un, " Jimmy the Fairy" the old 'un: the young 'un gave him a *stun,* and left him *stannin.*"  He also had it from Harry Brown, whose sister is married to Tom Jones, the head-lad's brother; and Jack Robinson, another friend of his, had it straight from the right quarter—from Nosey Jones, who gives Captain Noddle (his pet pupil) lessons in the manly art of self-defence. The Captain has horses in the same stable.

But what is that thing that has just made its appearance, looking like a cross between a ringtailed monkey and a gorilla?  Sammy whispers,—

" Mr. Tadpole, a friend of Mr. Squeezem's.  Another ' chip of the block,' eh ?"

" Bless my heart! why he looks as if he had been fed all his life on parchment-slips and scaling-wax !"

> [*Taddy surveys the company, and discovering his friend, esconces himself.*

" Well, Squeeze, how are you ?"

" Poorly."

" You've lost, eh ?"

" Of course I have."

" Oh! you'd back your friend Polish's nag."

" Rather.  Bad work, Taddy! bad work! discounting better than backing horses, especially such infernal hot 'uns. But come, cheer up, old cock! there's a good time coming: *They* may crow *now,* but *we* will raise the discount—eh ?"

" A good time coming, indeed!  ' Live horse and get grass.'  How do, Sam?  Quite well ?"

" Jolly, Mr. T.; how's yourself ?"

" Tol-lol.  Waiter !"

" Yessir."

" Some gin and walnuts."

" Got no walnuts, sir."

" Then go and get them, you muff! Do you suppose I can take my gin without nuts?"

> [*John retires, muttering,*—" He's had too much already."

Here one of the company remarks to another,— " That old chap is a regular ' out-and-outer;' if ever you want *any law done*, he'd go down a chimney after a chap to get it." " He's a rum 'un to look at, however" [*after a good look*]. " I've seen some very decent discounting lawyers, and good-looking fellows, too; but I should have taken that 'un for a tailor."

> [*In marches a party in sporting costume*—John Jogabout, *waiter from " Fair Rosamond's Bower," Richmond; seats himself beside* Sam.

" Well, your nag won, John?"

" Yes. Touched a little fifty ' quid.'"

" How did *you* get hold of this ' good thing?'"

" You see, I looks after the private rooms, where gents and their ladies comes for a week. One sporting gent, who often comes, told me it was a good thing. His lady lost her lapdog—such a beauty! and she gave me five pounds for finding it: when they were leaving he told me the Huntsman for Liverpool, too, he was with us at the time."

" Waiter!"                    [*From three quarters.*

" Here, Stupo!" (*from* Tad.)   " Got those nuts, yet?"

" All right, sir; coming."

" So is Christmas. What a stupid fellow your waiter

is, Sam! Here, bring some more gin. What are you doing, Squeeze?"

"I've just been thinking we ought to cut this confounded backing, and take to laying; it don't pay: one would want the Bank of England at his back. Let's make a book between us."

"A good idea! We'll talk it over."

> [*Tightfit's party is becoming rather noisy; one in particular very tight. The room filled with smoke. A voice from a corner exclaims,—*]

"Let's have some air, for Heaven's sake! Open that window over there, Sam." [*Just over Tad's head.*]

"Most emphatically, No! Not if *I* know it. No idea of sitting in a draft."

> [*Hates drafts, except of gin, or on the bank. Taps his snuff-box, takes a pinch, sneezes. coughs, and uses his pocket-handkerchief.*]

"Here come the nuts, Taddy, my boy."

"That's better. Where are the crackers, Stupo? Do you fancy I can break them with my teeth? Sam, Sam, this is dreadful!"

"Why, we shall smother! I'll stand this infernal smoke no longer. Come, I'll have that window open, or know for why."

> [*After a deal of persuasion Sam effects an arrangement ; the parties change seats.*]

"Capital cigar that of yours, Mr. Tadpole," (*remarks one of the new company*). "Might I beg one?"

"You may, if you like; but not from me. I got a present of a box from a client: he told me they were prime, and to keep them for my own smoking. I know him to be a good judge, and mean to take his advice."

" Thank you for nothing."

" Don't mention it."

> [*The Tightfits are becoming uproarious. One sings "* Old King Cole.*" Squeezem and Taddy keep on talking ; Tad pulling away at the gin. Squeeze, in a dialogue on the imprudence of professional men meddling in such speculations, asks Tommy "if he understands Latin ?"*

The noisy Tightfit man, at the conclusion of the song, rattles away with a large tankard on the table for the waiter, which causes Tad to jump from the effects of gin, at the moment Squeeze has said to him, " *Ne sutor ultra crepidam,* Taddy." Up jumps the tankard man, and asks Tommy " how he *dare* make such *impident* remarks ?"

" What do you mean, fellow? Sit down and draw yourself to an anchor. I did not address you."

" You *insinivated* that I made the soot come down the chimney."

> [*Sammy starts up, makes peace, assures him of what he verily believes—*" The gents was only talking French; it's all right: nothing meant." *All becomes serene—A devil of a row in the outer bar.*

" What's the matter ?" asks Sam.

" Nothing, sir; it's settled. Ned Greenham the costermonger, and Bill Jenkins the tout, had a few words. Bill told him to back ' Miss Cruiser ' for the Oaks, and demanded his fees : Ned said it was ' Matilda Tightwaist.' Bill is right, for I heard him say myself she couldn't stay; he'd cat her if she won : drawn too fine ;

was on the go; done too much work; had a bad
night, and looked as if she was dragged through a
hedge backwards, and hadn't eaten a feed of corn for a
month."

> [*Tommy's health, with three times three,* " For
> he's a jolly good fellow;" "Touch him with
> the crowbar," (*another,* "With the poker.")
> *Tommy becomes glorious; stands a round;
> sings* "Cheer, boys, cheer;" "We won't go
> home till morning :" *in the middle of which
> the Knight of the Napkin announces that a
> lady in a cab requires Mr. Tadpole's im-
> mediate attendance.* "Cannot be him; must
> be Mrs. S." "No, it's Mr. T.;" *who,
> with some difficulty, is bundled into the cab
> with Mrs. T. The company, after drink-
> ing the health of Samuel Shandigaff, and
> many happy returns of the Derby day, dis-
> perse to their domiciles.*

The Derby week in London affords many opportunities
of witnessing amusing scenes. A rather funny one took
place at a certain well-known and long-established hotel,
frequented principally by elderly wealthy gentlemen and
rich merchants. One of its patrons was a middle-aged
bachelor, who, having nothing else to occupy his attention,
devoted his talents towards proposing amendments to
resolutions at public meetings. He was most eccentric
in his habits; of hasty temper; should have first attendance with everything—first of any dish; newspapers,
which he appeared to spell from beginning to end, and
monopolise, to the very great annoyance of others. His
attire it would be neither necessary nor courteous to refer

to, further than to state that it was a light green coat; collar about six inches deep (partly sheltered by very long whiskers, and moustache of immense size); tails extremely long, and in the good old style, looking as if built in the reign of Queen Anne, with a superfluity of brass buttons from top to bottom, in which he appeared to live exclusively, and likely to die, if not to sleep; a Paul-Pry umbrella winter and summer—from which fact he was known by the name of " Billy Button."

Having been in company with a friend of mine one evening during the week, two others entered. The one a sporting lawyer; the other had been a captain in a dragoon regiment—we will call him " Jack Rollicker;" remarkably good-looking, with a wonderful flow of spirits, most humorous manners, which, coupled with being a most liberal, generous, and fine-hearted fellow, had made him a general favourite. His ideas of the value of money bore a striking contrast to those universally entertained, he having spent the greater portion of a fine fortune. Having just returned from the races, he appeared at least in as high spirits as usual; and previous to sitting down, in a jovial manner was relating some of the incidents of the day's racing. Mr. B. had been reading the paper with the additional aid of a wax candle, sheltered from the gas above by a large piece of paper, which he held in the other hand. Having dropped the paper and fixed his eyes most intently on Jack from under his specs, the latter surveyed him, and at once struck with his peculiar *tout-ensemble*, bowed courteously, congratulating him upon his good fortune. Mr. B., in an astonished and rather irritable manner, replied,—

" You are mistaken."

" Met you at Limmer's, have I not ?   Heard you had won a good stake on the Derby."

" Never, sir ! never !   Never there. And as for stakes, I never eat them."

" No, no : I mean, that you won a large sum of money on the race for the Derby."

" Never was at a race, sir ! never ! You have evidently mistaken me for another person. Never was at Derby, and detest horseracing."

" Extremely sorry, sir.   Must be a mistake."

*[Dinner over, various parties taking their wine at tables around; and in the centre of the room Mr. B., still reading the newspaper. The sporting lawyer remarked that he expected a blaze shortly, that the paper was at times absolutely touching the candle between the shade.   He had scarcely uttered the words when away it went like a balloon on fire, landing amongst a few old gentlemen next table, who fled in all directions, Billy shouting, "Waiter ! waiter ! Water ! water !"*

*A discussion having ensued upon the subject of racing, various parties in the usual way had joined, while several old folk fiddled with their large bunches of seals, and sipped their wine; Jack surveying them, and selecting which of them he should fancy as likely to be possessed of most of the " sinews of war," chose one who had been speaking to himself—as he remarked, probably about consols and railway shares.*

The sporting lawyer, in advocating the turf, remarked,

" It was the best school to make a man of the world;
that everybody was running after money; it afforded
opportunities of seeing life in most classes of society;
that nothing whatever could be justly or fairly urged
against it."

An elderly gentleman here remarked, that " he should
be very sorry to have his son educated there, or know
anything about it ; that it was a most dangerous
gambling pursuit, a speculation only fit for rogues or
fools. He fancied Harrow would be a more desirable
school for youth."

[*Billy here looked from under his specs,
and smiled assent.*

" True ; but if I wished likewise to enable him to
*plough* his way through the world, I should give him a
slight knowledge of the turf: even if he were to become
a bishop or a judge, it could do him no harm."

" Pshaw ! Judges and bishops totally discountenance
it. Nothing but gambling !"

" Decidedly not. Many of them would enjoy a good
race, and like to have a little pecuniary interest into the
bargain, to add to the excitement. Where is the profession or calling that is not in some way, or to some extent, a speculation ? The world, sir, is money, money —
a race after money."

" No argument, sir ; not a particle ! I 've known
many men who have lost considerably in turf pursuits,
and been obliged to come to us, when raising money, to
pay such liabilities. All gambling dreadful ! I speak
from experience."

" Very good of you, sir, to accommodate them, and of
them to pay their engagements. They are remarkable for

their strict sense of honour and punctuality. Pray, sir, in case I or my friends should require your kind assistance, would you favour me with the nature of your profession, or calling ?"

"I happen, sir, to be, in addition to other matters of business unnecessary to mention, a Director of an Insurance Company."

"Oh, indeed! What are Insurance Companies ?"

"A company, sir, composed of a number of proprietors ——"

"A number of gamblers, you mean, according to your own account. I fancy I have you *safe* now; presume you sometimes lay against a dead 'un."

"Do not understand your terms, sir. *We* are a consolidated fund, capital one million sterling; and as to *my* being safe, I am safe for thirty thousand pounds, sir. What do you think of that ?"

"Wish I had it now," says Jack.

        *[Billy looks from under his specs at him.*

"But still you must be sometimes hard hit. Suppose you lay five thousand to one hundred against a young 'un; I think those are about the odds; for instance, a fast 'un like my friend here, and 'his goose is cooked' immediately afterwards, how do you square your book ? — how do you hedge ? Get it out of the old 'uns, I presume ?"

"Really, your terms are Greek to me; and as for your figures, they are fabulous."

"Pooh! pooh!" says Jack; "mere fractions!"

"Suppose you lay twenty monkeys to a pony against a house; it takes fire in a fortnight and burns down, perhaps with a few of your patrons inside: how do you get out of the fire ? That takes some getting out of, don't it ?

Defend the action, and plead the owner was the incendiary, eh? I'm a bit of a lawyer, you see. As for gambling, you are for ever getting up speculations; the next will probably be an earthquake and balloon for the shareholders to escape by."

" Never was in a house on fire, sir, thank God, and hope I never shall be. Don't feel disposed to continue this line of argument. All Greek to me — all Greek; and as for lawyers, not particularly fond of them."

" Neither am I, I assure you."

" Suppose you enter a *Nolle prosequi ?*"

" A *nolle* what? You seem to have a peculiar mixture of languages, as well as professions."

" You don't appear, sir, to have been at Harrow, then? Neither have I been, although often at Cambridge. A great favourite of mine that country."

" Capital! jolly!" adds Jack; " one of the most glorious spots in the world!"

> [*Mr. B. calls for second edition of* " The Sun."

" Hope you won't make a second edition of it, sir," adds Jack.

> [*A venerable old gentleman, who had escaped the fire-balloon, gently drawing his seat towards Jack, remarks,* " I trust not;" *requested to join the table: does so, addressing the sporting lawyer,* —

" Happy to hear, sir, you and your friends speak so highly of Cambridge. I have there, at this moment, my only son; a most promising youth."

> [*Jack calls for a bottle of port (his favourite wine), and asks Billy* " what they are doing on the ' Oaks.'"

" Don't know anything about it ; you have evidently mistaken me for another party."

" Would you favour me with a peep for a second ?"

" The waiter, I dare say, will find you another."

[*Waiter brings fresh bottle.*

" Only take in one of the second edition, sir." *Whispers :* "' Miss Cruiser's' come a cracker; two to one taken freely."

" That'll do," says Jack ; " I'm on at ten."

[*Tips him a crown ; waiter slips him* " Locket's Circular," *and adds,—*

" He gives ' Miss Cruiser' to win, and ' Pinsticker' for a place."

" Now, sir," adds Jack, "perhaps you would prefer a glass of Chambertin ; but here is a capital glass of old port. A prime glass of wine as any connoisseur need wish for."

[*Tastes it : after the usual peep, turn, and smack,—*

" A prime glass of wine."

" Yes : they keep everything very good here."

" And very moderate. Bless you, my dear sir ! I have patronised this house these twenty years; hardly miss a day, except when I run down to see my boy at Cambridge."

" Ah, the very name of Cambridge puts life into me," remarks the lawyer, helping himself. " Pray excuse me asking if you have yet made up your mind as to what profession you mean him for ?"

" My dear sir ! Most happy to tell you. *His* mind is made up, long ago, upon that point."

" The army ?"

" Right sir, right!" says Jack.

" The army, sir, and nothing but the army for him.
I assure you he sometimes acts in private theatricals, and
always assumes the part of an officer in full uniform.
Wonderful taste for the army! He is already growing
a moustache—ha! ha! ha!"

" Perfectly right. It is the true school to make a perfect man of the world—next to the turf. A young man,
especially about to join the army, should always have a
knowledge of what he is likely to have to contend against.
I'm a great advocate for young men knowing as much of
the world as possible, before they embark in any pursuits
wherein their intellects are likely to be called into action.
Nothing like it, depend upon it. They frequently have to
pay too dearly for their experience. Nothing like an
officer knowing the enemy. Good idea of yours, sending
him to Cambridge."

" Then you think I could not have selected a better
school ?"

" Decidedly not. He will have the double advantage.
I have a son, but not at Cambridge; although some notion
of finishing him there. He has already taken first places
in classics, Greek, Latin, &c. I intend to instruct him
myself in the other rudiments of human nature, having
been plucked as clean as any green goose at Michaelmas;
fearing the young 'un might some day take after his
parent, in his fondness for racing, although by no
means so inclined at present: it sometimes runs in the
family."

" My Albert has never shown any inclination in that
way, although very fond of horses. He is most attentive
to his studies."

" Ah! no telling the moment. And in order to guard against danger and prepare him to meet his foe, The World, in this battle for money in so many shapes, my son is already pretty well up in the various requisites. I shall back him to beat any lad under sixteen at all the classics —to speak French, Italian, German—and what so many of those highly-finished scholars are deficient in, *his own language*—to tell the pedigree of any horse in 'The Racing Calendar,' and wind up by running four miles over the Beacon Course at Newmarket for a piece of plate, value one hundred guineas — or two hundred either—to be handed to the winner, and stand a dinner for six, if any sanguine parent will make the match."

" Bless my soul! what a wonder he must be!"

" Yes, sir; so his grandmother thinks. He is a very promising and good-looking youngster; but rather tall and overgrown at present: but will thicken with time and training, I expect."

" Pray for what profession do you intend him?"

" Have not quite made up my mind. Had some idea of the Church; but, upon mentioning it to him, he jokingly remarked, ' he was better suited for the steeple.' Probably he meant steeple-chasing. No telling the moment."

" Is it not a pity to encourage him in those ideas?"

" By no means *encourage* him; on the contrary: but ' blood will tell.' Every day expect to find him show some speed in that direction. His grandfather, although always opposed to absolute proprietorship of racehorses, was remarkably fond of a good race, and of horses. He ascertained that I had some under the rose, and called me, in angry terms, a horseracing scamp; but having accidentally heard of my success, to my astonishment almost embraced me! I occasionally figured in silk, *sub rosâ*—a second Pierce O'Hara of '98, preferring 'The Racing Calendar' and pigskin to 'Blackstone's Commentaries' and parchment. The Race *versus* The Case."

    [*Billy looks under his specs; a long look.*

"Dear me! It would be the death of me if my Albert were to attempt such a thing."

"He's all right enough," says Jack. "Does he fancy The Plungers, or The Mudcrushers? He is certain to ride in the Garrison races."

"Have heard him speak of the Blues and Buffs, and others; but never heard him mention either of those regiments."

"They're a very nice mixture, I assure you. Blue body and buff sleeves would do nicely; and lucky, too, 'Faugh-a-Ballagh's' Sellinger colours. I backed him for a monkey. Won't you help yourself, sir, and pass the bottle?—not a headache in a hogshead of it. Waiter, a fresh bottle of port."

    [*Billy drops the paper; looks first at Jack, then*
     *at his watch—almost his regular hour.*

"I think you said you had an idea of finishing your son at Cambridge, did you not? You prefer it to Oxford?"

"Have some notion of it; but, unfortunately, my financial department is not at present in a very flourishing condition, owing to various causes."

"Ah! losses on the turf, I presume."

"By no means. Quite the contrary, I assure you. Although not much of a speculator or money-hunter, never a tuft-hunter, and little of a fox-hunter, but rather

inclined to hunt after other game, such as pheasants; fond
of greyhounds (once beat the celebrated Father Tom's
'Lady Harkaway,' and won a gold cup to his mortification), poodle-dogs, toy terriers, an odd shy at hazard, and
other little innocent amusements, which cost money: but,
worst of all, entertaining and accommodating my numerous friends. Spent, sir — rather, squandered — many
thousands; part of which had been made upon the turf."

[*Billy drops the paper, looks at his watch, and
commits himself to the care of Morpheus.*

" Gracious me! Am I to understand that a clergyman kept racing-dogs?"

" Certainly! Some of the best that ever followed a
hare: 'Doctor Syntax' and 'Lady Harkaway' about two
of the best I ever met. I kept thirty greyhounds at
one time, and lost a fortune by one, in a most curious
way. Too long a story, sir, to tell you. Suffice it to say,
he broke into the house; jumped upon the table after
some bones of fowl; upset a quantity of things, which led
to a question of whether my parent would shoot the lot;
and drove away an old dying relative and godfather, who
forthwith burned his will, which had been made all in my
favour, and died a week afterwards. He was an old
miser. My father, sir, many a time cursed the greyhounds,
and shipped the lot. I once knew a parson (a client of
my own) who told me he detested horseracing. I found
him, upon a subsequent occasion, in the centre of the ring,
endeavouring to become initiated in the Eleusinian mys-
teries of the turf."

" Very curious case. I should not be fond of greyhounds under such reflections."

" Next to racehorses, sir, the best fun, to my fancy."

" Greyhounds be hanged!" says Jack. "No music or fun, those see-dogs, pot-dogs."

                         [*Helps himself and passes the bottle.*

" Why do you prefer Cambridge to Oxford?"

" I run down periodically, about four times a-year, and spend a week in the neighbourhood."

              [*Here a fresh bottle of wine is placed on the table; Jack helps the old gentleman, then himself, and passes it, remarking,—*

" Capital partridge-shooting, and Newmarket close by; celebrated for its delicious ——"

" Newmarket! Bless my soul and body! Newmarket close by! I never dreamt of that astounding fact!"

" Of course," adds Jack. "Glorious place! You should make it your business to run down there, see your son — kill two birds with one stone, as they say: he could give you a seat over in his carriage. They are always racing there, from New-year's Day to Christmas Eve; from morning till night: sometimes in a fog. Enjoy yourself very much — capital sport — add twenty years to your life — bracing breezes!"

" Why, sir, my son has no carriage yet! his father had not one until he earned it. But do you really mean to say that place is so very near?"

" True bill, sir; one of the jolliest spots under the sun. You can win twenty or thirty thousand as easy as snap your finger — aye, fifty thousand, if you like."

" Oh, dear!" (*with a deep sigh.*) " Oh, dear! Albert never mentioned anything of having visited races, except the boat-race. He formed one of the Cambridge crew, who were victorious."

" The very thing that will give him a taste for racing.

Why, all the lads join in a fly or a drag, and run over there in a jiffy; or the train would drop you there in a few minutes. It's the finest place in the world. Talk of America, Australia, or the gold diggings! why, they are copper-mines compared to it — for speculation: you might make a fortune in five minutes there, and have glorious fun into the bargain; whereas you might have to slave all your life for a paltry twenty or thirty thousand pounds elsewhere."

"*A paltry twenty or thirty thousand!* It's very easy to talk of snapping, but it might be snapped from me. I assure you, my dear sir, I have spent all my life, now sixty-two——"

"Sixty-two!" interrupts Jack. "Why, sir, you don't look over forty. A few fresheners over the plains of New-market would make you as fresh as a kitten. You would become fond of the fun — would often run down and see Albert; take a run over, and have a little shooting. Capital partridge shooting."

"I was just about to remark, that I have been all my life a most industrious adherent to business, and found it extremely difficult, what with losses in speculations, rail-way and other shares, to realise a sum very far short of a plum."

"A plum!" says Jack. "Oh, Albert! lucky dog! Couldn't have prettier or more lucky colours. Buff and blue, eh?" [*Helping himself*.

"It is very difficult to make money, but sometimes people find it more so to keep it."

"I believe you, sir. I never made any, nor my father before me; but whatever he couldn't drive through I finished in double-quick pace."

"You said," (*addressing the sporting lawyer*) "that things of the kind sometimes run in families; I should fancy, on the same principle that, if the parent were industriously inclined, and had, for instance, made his own money, and displayed a knowledge of its value, it would be natural to suppose the son would be thus inclined also."

"A great deal would depend on circumstances. They run in all shapes, and sometimes throw back to the grand-sire, for instance, if he had been fond of cock-fighting."

"Cock-fighting! Bless me, sir, my father abhorred it!"

"Oh, I do not refer to him especially, or to cock-fighting in particular: but as an example, it's certain to break out in some form, even if it escapes a generation. The blood will tell, sir, I assure you; like the gout, for instance."

"Gout! Thank God, none of my family ever had it."

"No, no. I merely quote it as an example."

"If they had, rest assured I would not dream of touching this delicious glass of port."

"Bless you!" says Jack, "I have it now, at this moment; and here I am, helping myself."

"Thank God, it's not contagious!"

"No, no. But I was remarking, that as to mania for spending money, it is a remarkable fact that young men, whose parents have evinced an over-fondness for money, frequently turn out the contrary, and generally select the turf as their favourite pastime."

"Suppose I remove my Albert to-morrow, and shift his quarters to Oxford, for instance?"

"Bless you, he would never be fit to plough through the ranks in life, as the Cambridge education will leave him. Let him remain where he is, my dear sir. If you check the natural impulse you only increase the mania, if it exists; you may rest assured he has been slightly inoculated already. How long has he been there?"

"Nearly two years."

"It's all over but the shouting. Buff body and blue sleeves," says Jack, rather elated, helping himself, and requesting the old gentleman to follow suit.

"How do you mean? No hope for him?"

"None."

"None, whatever," says Jack.

"But he may be ordered to India?"

"They race there, also."

"What! in India?"

"Of course they do. They race everywhere. Everybody races, especially after money."

"Happy to say I never did."

"But, my dear sir," adds Jack, "if your son is fond of the army, he is certain to be fond of racing, and better let him have his fling. He must get ready for the Garrison races."

An elderly gentleman at next table remarks,—"Pardon me, I have been rather amused, but is it exaggeration or joke when you say one can realise such large sums of money upon the turf, in so short a time?"

"Exaggerate! Why, let me see. It takes about four minutes to run the Derby. You can win forty or fifty thousand pounds on it;—aye, nearer to a hundred thousand, if you have a mind."

"Yes, with proper pluck to put it down," says Jack. [*Helping himself, and throwing off with a gusto.*

"If you chose to make a certainty, without risk, of

say thirty or forty thousand, by going to the Insurance Office——."

[*Here the Director turns upon his chair, and remarks,—*

"I beg, sir, you will not refer to Insurance Companies in a sarcastic manner."

"Not referring to you, I assure you: nothing personal, sir."

"If you, for instance, back a horse for one thousand pounds ——"

"That's the sum," says Jack; "say two."

"You get, suppose, to your thousand, fifty to one. The consequence which invariably follows is, that from the fact of your having backed that particular animal for that sum, various other parties will follow suit; the animal comes to perhaps ten to one, but certainly to much less odds. You can then turn round and hedge; or, in other words, save your own money by going to the Insurance Office, as it is termed; you then stand to win, in the event of the horse's success, forty thousand pounds, without a penny risk."

"By Jove!"

Jack aside, says to his friend, "I've known them to be backed for many a thousand, and get worse favourites. Still, I think we have a convert here." (*Helps his friend and himself, and aloud, in his usual voice, adds,*) — "I always go the whole hog—stand it out, unless at very tempting odds, or to oblige a friend, or the party who has laid the odds, and thus help him out: further than that, no Insurance Offices for me. If you don't put it down you cannot expect to take it up. P. put down, T. take up, is my motto; that's the style."

[*Director casts a glance, and smiles.*

" Touching the mode of realising without risk, it strikes me the hedging system, or insurance plan, is the prudent course, and the one I should select. I see my way clearly, with one difficulty, and a most important one."

> [*The cavalier's parent shakes his head, and anxiously awaits the reply to that poser, as he expects it to prove.*

" How about the payment ?"

" Payment ! why the best and most punctually paid money in the world — to the day, sir ! Their word is their bond. The sportsman will pay to his last guinea; the most humble of them will do so. Occasionally a slight panic, as in any other affairs, may occur, and cause a " temporary inconvenience" to a few, but they invariably come to time afterwards. Accidents will happen in the best-regulated families, and none more indulgent to each other under such circumstances. Fine-hearted fellows, sir, as any in the world."

" None like them," adds Jack, helping himself, and passing the bottle ; " always foremost in any emergency, where their fellow-creatures require aid, and yet the most abused men living, by some who do not understand the true state of the case. The first and highest-minded men living are on the turf, sir, I assure you."

" Really," remarks the cavalier's parent, " it seems plain enough there cannot be much harm done, with prudence."

" None at all, with prudence," says Jack. " The Oaks will be run for on Friday. Send for Albert, sir, and take a run down ; the ladies' day — he will be delighted. Be sure and tell him to back ' Miss Cruiser' for a monkey."

" Well, bless me ! it is considerably beyond my usual time, you have so agreeably entertained me. It strikes me there is a great deal more said against the turf than it really merits. I begin not to feel so apprehensive of dangerous results with Albert."

*[Puts on his coat and bids adieu.*

" Give him the proper education, sir, for a man of the world, in time ; never can tell the moment he may require it, especially in the army."

" This appears a very clear case," remarks the elderly gentleman. " In my railway and other speculations I frequently get a severe shaking, and see no reason why——"

*[Jack gives his friend a touch under the table ; the friend in return gives him one, which makes him almost faint with the gout.*

"——I should not merely speculate a little at first, as a mere trial."

*[Jack jumps up, half in pain, and, unable to restrain his desire to laugh, asks,—*

" Who's for a cigar ?  Any one in the smoking-room, waiter ?"

" Yessir, several gentlemen ; some strangers for the races from different parts."

This favourite and national pastime presents opportunities of viewing various specimens of mankind, from the highest to the lowest members of society.  Nowhere are the world's farces, its changes, chances, and vicissitudes, more wonderfully brought to light.  It is a " world of wonders" in itself.  We there see the finest representatives of nobility, and amongst them the finest samples of mankind; not alone noble by name but by nature;* men whose *maintien* (a word borrowed for the occasion, to be in the fashion. If I were to win the Derby, or be made an M.P., I could not tell the exact meaning of it; and as to pronouncing it, nobody but a Frenchman or Yankee could do so properly) and affable manner gain and secure that respect in which they are certain to be held by all well-thinking men. They bear a most striking contrast to "tuft-hunters," who, in the eyes of their fellowmen, are so deservedly looked upon with contempt, and by none in reality more than by the nobility themselves, one touch of the hem of whose garment converts them — in their own opinions — into aristocrats. There are none more certain than they to be at their post, always on the look-out to pick up any crumbs which may fall from their master's table, who sometimes tolerate them as a necessary nuisance, as they generally act the part of the " boot-jack:" in their toadying anxiety to make themselves appear in the eyes of others what they in reality have no pretensions to, further than what may arise from having squandered an amount of money, (accumulated by their more industrious and less presuming ancestors,) and having assumed towards their equals, and frequently their superiors, an air of arrogance and impudence, amusing as it is ridiculous; and who, though disclaiming connexion with the democratic drain, when " pushed from their pedestal," and when reduced in their financial department, are quite as ready to " bend" *for profit* in any

* " 'Tis not purple and gold that ennoble the man,
　　Nor the baubles the vulgar revere;
　'Tis the heart that can feel, 'tis the mind that can span,
　　'Tis the soul that no danger can fear."

H

respect, and to endeavour to persuade the credulous to a belief of the justice of their pretensions, by the assurance that the Duke of Dupes, Marquis of Muffs, or Sir Simon Swallowall, are their most intimate friends. Such personages may be best described in the words of Moore : —

> " Beside him place the God of Wit,
>  Before him Beauty's rosiest girls,
> Apollo for a Star he 'd quit,
>  And Love's own sister for an Earl's.
>
> Did niggard fate no peers afford,
>  He'd take, of course, to peers' relations;
> And, rather than not sport a Lord,
>  Put up with e'en the last creations.
>
> Even Irish names, could he but tag 'em
>  With ' Lord' and ' Duke,' were sweet to call;
> And, at a pinch, Lord Ballyraggum
>  Was better than no Lord at all."

Another of the enemy's scouts, always found hovering about on the look-out for prey, is the officious, accommodating usurer, sometimes with the beneficial addition of " lawyer" to his title; who, with the assistance of his satellites, like the shark following the ship, waiting for the corpse to be cast overboard, never leaves the unfortunate victim while a spark of life remains in the shape of a sovereign to extract, to satisfy his wolfish appetite. *He* also finds it a necessary acquirement towards carrying out his enterprise to assume a certain amount of consummate assurance; and should any hapless scion of a noble house have been at any time compelled to seek his suffrages — should his name be on the *tapis* in any company, — he is invariably mentioned in familiar terms by him, as if they had been reared like foster-brothers together all their lives;

and probably a seat in the Senate, or liberty to lounge on the cushions of the Carlton or Arlington, would suit their aspirations and their purposes: where, no doubt, they would resemble a living fat turtle, which I saw some time ago, strapped and sitting up amongst the passengers on the top of a 'bus, in the neighbourhood of the latter club, looking down with a happy air of consequence on the amused spectators, whatever may have been his meditations as to his position, or those of his fellow-passengers as to their wishes. Those worthies sometimes hover about racing establishments, like birds of ill-omen, wherever they fancy their temporary assistance might be required by master or man; and occasionally place their eggs in more baskets than one, in the hope that they may produce more fledgelings. When they have once got their victims in their claws they turn their weapons to the best advantage, by sometimes extracting information *in terrorem* from those whose position might be compared to that of Damocles, when he found himself seated beneath the sword suspended from a horse-hair.

Of all the pests that ever frequented the racecourse, and the one most to be shunned by young and inexperienced men, is the over-dressed, polished, card-sharping practitioner, whose sole aim is to become acquainted with youths whose fondness for sport may lead them to follow turf pursuits. Once the opportunity presents itself, the die is cast. Talk of thimble-riggers! the world knows what must be expected at their hands. But woe to the victims of gentlemen card-sharpers and flat-catchers, who will lend themselves to any foul play!

Let the best judge that ever lived possess the best horses that ever galloped, let him have the wealth of a

Crœsus, and then he may not be successful. On the other hand, a party may have moderate animals, and be a comparative neophyte in such matters, and yet he may be most fortunate. Why? Simply because the owner, in many instances, has no more to do with the animals than the child unborn, further than purchasing and paying their expenses. So much depends upon the principle and integrity of the parties in whose hands it may have been his fate to have placed himself; some of whose system of management consists in their talent to put thousands in their own pockets, and leave owners to shift for themselves, under the flag of "honour and glory." Many of such managers, as they term themselves, know as much about a racehorse as did schoolmaster Squeers' pupil —that "*it* was a beast." Yet some of these individuals strut about with a bloated air of purse-proud consequence —the thoroughfare of Piccadilly hardly wide enough for them—chuckling over their unexpected treasures, which they had probably gained through the knowledge and experience of others, whose brains they had sucked, like the insects that prey on the brain of the elk, till his very last sigh. Nor should owners fancy, that because they have good horses they must, as a natural consequence, be successful. It by no means follows. On the contrary, many men have lost considerably by good horses, and the largest sums most frequently upon them. The sanguine disposition of some persons leads them into the belief that their animals are invincible, forgetting that others have good horses likewise; without which prejudice, however, racing would soon be at an end. Still, it is easy to overcome the dangerous effects thus produced by studying the common dictates of reason and prudence, and adopting the safe system of hedging, or saving their own money, and thus going to "the Insurance Office:" for it is natural to assume (although it unfortunately is not always the case) that the owner *ought* to have the *first* information of the merits of his own animals, and make use of it by having the first run of the money market. One of the great mistakes frequently made is, that they do not at the proper time back their horses for an amount adequate to reimburse the heavy outlay which attends the keeping of horses; and when they find them fit and well, do not put down the pieces of gold in preference to continually risking money under other circumstances, and upon shallow chances.

Indeed, so long as men follow such a pursuit, no matter how judiciously they may act, there is a tide of success, or the reverse—a sort of fatality attending such speculations, accompanied with a run of either good or bad luck; for those who have reaped rich harvests were not often the owners of the best horses, nor the best judges of how to manage or place them, which establishes the fallacy of believing that it is at best anything but a most uncertain adventure. Still, there is no speculation by which a fortune can be realised in so short a period, and with such comparative certainty; provided the follower is a sound judge of the animal, and knows when and where to engage him, and provided he has coolness and steadiness, never being over-sanguine, or holding others in too slight estimation. But it would, in some instances, almost lead one to become a fatalist, when we witness the run of ill-luck which attends some of the most zealous patrons, as if they were doomed never to realise their expectations and ambition; and yet some of these parties are the most straightforward, and in every respect independent adherents. Others bear away the palm with, comparatively speaking, no pretensions to cope with them in any point of view. It seems strange, however, that if there really be such a thing as luck, or fate (an attempt to arrive at the truth or meaning of which renders one very much like a man endeavouring to discover the perpetual motion), that it should follow any person more upon the turf than in other pursuits. Yet Dame Fortune is nowhere " more *female*" than there; dealing out her favours in a most whimsical and unjust manner, dispensing them in a most lavish manner to some of her votaries, withholding all from others, who might well be compared to the Irishman, who thus described his luck when writing to his wife,—

" Bad luck has two handles, dear Judy, they say;
But mine has *both* handles turned on the wrong way."

The turf is a contest between men, whose extreme confidence in their individual superiority over each other, in the art and science of horseracing, renders it, according to their own belief, not only absolutely possible, but highly probable, for each to monopolize the greatest share, if not the whole, of a complement unlikely to satisfy any, and totally inadequate to satisfy all; which fact is its life. The food with which it is supplied emanates from fresh sources, continually flowing in the shape of young recruits, who join the ranks for the purpose of amusement and instruction, some of whom promise to make good generals when they join, and really turn out most able officers. Still, the majority of the number pay most liberally for having mistaken their profession —

a fact which, taken in a philosophical point of view, is not only a happy circumstance for the turf, but may naturally and fairly, according to the dictates or rules of human nature, be looked upon in the light, that having selected it as their favourite pursuit and that one most suitable to their tastes, it is as well it should be so, inasmuch as they might have chosen another, equally dangerous and with less recreation. An additional consolation, in a pecuniary point of view, is, that they have the gratification of knowing they have been instrumental, if unsuccessful themselves, in maintaining their favourite princely pastime in its splendour, and prevented its followers sharing the fate of the Kilkenny cats.

The turf, upon the accession of a few young, wealthy patrons, might be compared to a fancy-fish vase upon the addition of fresh water. How they frisk about! the young 'uns in particular, the old 'uns likewise, with open gullet, becoming more enlivened, and with more steady wag of the tail floating about, generally after the large crumbs, although the young 'uns show more speed, and make more numerous snaps at the smaller ones.

Taking it in the light *solely* of a pecuniary enterprise, there have been many of its oldest patrons who have tried the experiment of keeping an *extensive* stud, with but one result. Some have kept something like one hundred horses. There is the grand mistake. In the first place, the everlasting drain upon the exchequer ; the wear and tear of the very animals themselves, in loss of price and value, through breaking down or other causes ; the impossibility of becoming owner of many good, or even middling ones ; the disappointments attending even those that may turn out well, setting aside the positive certainty, that some of the most promising and high-priced ones will prove worthless, and run away with the profits of their betters; coupled with the fact that the competition for prizes run for in these days is so great, and the amount of stakes contended for so disproportionate; all combined make it evident that one would require the purse of a Rothschild, the patience of a Job, and the temper of a Socrates, to carry on a large establishment with anything like reasonable success, if he relies *solely* on the amount of *stakes* likely to be won: for although he may have what is termed "a good year," or a "run of luck," the mere stakes cannot possibly meet the enormous outlay attending *a very large* establishment. It may be that the owner speculates largely: and then, in nine cases out of ten, his temporary success carries him on, and leads him to repeat his heavy investments, when it becomes a mere question of time as to when the plunging powder-magazine will burst, the tide turn, and carry away not only the previous gain, but a great deal more along with it. The great difficulty is to know when and where to stop, and never to " halloo until out of the wood."

On the other hand, it would be absurd to deny that there are plenty who have done wonders with a small stud, where judgment, not only in the animal, but moral courage *forthwith* to get rid of bad ones at *any sacrifice*, has carried the owners through. *They* have plenty to contend *against*, and plenty to contend *for*, supplied by those who make the mistakes referred to. *There* is the value, which is not obtained, and cannot be gained, through *numbers*, but through the instrumentality of *good* and well-directed instruments, used with caution by experienced hands. Judgment will always beat money in the long run. But those by whom fortunes are principally made, and can surely, and, comparatively speaking, with little difficulty, be realised, are those who speculate, not upon the anatomy of horseflesh, or their qualities, or breeding. The better they are, the better they like them, and the more profitable they are to them, for many reasons. They know nothing about them — the less, perhaps, the better for themselves; nor probably would they wish to know. Those who look not at the animal, but at their books and their figures, who diligently and anxiously seek the sanguine owners, and test their fondness and partiality for their respective favourites, the more the better. Men who think nothing of paying to the fortunate owner of a winner many thousands, after doing which, in many instances, they remain still large winners by his success, and sometimes find their troubles rewarded with the result of not having to pay any one of the parties who scientifically endeavoured to make a large "aperture" in their book. Such are the parties who, no doubt, honourably and fairly amass fortunes; and why not? Because they have not the millstone round their necks, in the shape of the heavy expenses referred to, besides stakes and forfeits; nor yet the perpetual worry and torment attending the keeping of racehorses, so frequently called out of their names. The news of a break-down or other mishap on the eve of a race, is of no deeper concern to them than being a profit to their purses. They are not uneasy when the postman delivers a letter in the well-known handwriting of the trainer, on the eve of a race, announcing to the owner: — "My Lord," "Sir," or "Dear Jack," (as the case may be), "I am sorry to inform you 'Tom the Devil' has caught a cold;" or "'Mrs. Bang-up' has broken down;" or perhaps the gratifying epistle may run thus:—"I tried the young ones this day, and am sorry to inform you they are all bad; but sincerely hope the next lot may be good." None of those relishes for breakfast await or belong to the legitimate bookmaker; he meets the admirers of the respective animals as they come, and deals with them as a matter of business and *figures* — not of *fancy*.

It appears nevertheless strange, why bookmakers should be so frequently singled out, not only as a target for the long shots of the backers, generally rejoicing in the name of the "gentlemen," but it would almost appear as part of their duty sometimes to bear the odium of those who may have missed the "bull's eye," or "thousand to ten," and failed to realise fortunes in their dealings with them. A Bookmaker may be called by that appellation, yet he is neither more nor less than a speculator, who thinks proper to devote his time and employ his capital in the pursuit he deems most suitable to his taste or talent, and naturally considers that he has a right to live by "the sweat of his *brain*, as well as by that of his *brow*." As an Insurance Company (and what is an Insurance Company, as that lawyer termed it, but a bookmaker?) risk *their* capital, so do bookmakers risk theirs. The one receives a premium on the conditions of *paying* a certain sum in the event of death, the other in the event of a certain event coming off. The difference in one respect being, that it sometimes happens the *sporting* insurer *receives* on a "*dead'un*." Some parties, unacquainted with the real nature of such subjects, frequently run away with very extravagant notions as to the position of bookmakers. Their transactions are ruled by the market, as others are on the Stock Exchange, where there is far more "gambling" than on the racecourse. Quite as many men have taken a liberty there, not only with shares, but with the disposal of their own lives in consequence, as in the other case. It matters not how humble a man's antecedents may have been, nor whether he may have other avocations to attend to (however inconsistent); if he thinks proper to add a branch thereto in the metallic line he has a perfect right so to do, and his bank-notes are quite as good and as acceptable, whether they odour of patchouli or pigtail. If he succeeds in his undertaking, the more credit is *due* to him in one respect, and the more he will *get* in the other, if he requires it.

What difference there can be between one man who *takes* odds and another who *lays* them, as far as constituting in itself respectability or morality, is a problem to be solved. The powerful influences of "palm paste" effect both to a certain extent; if more so in the one than in the other, it is on the part of the backer, who seeks to *gain more* by *risking less*. There can be little doubt on one point, however, that if bookmaking were less difficult or troublesome, more of the "gentlemen" would condescend to try their hands. Some who have done so, under the delusion that the science consisted merely in purchasing a betting-book with gilt edges, and a gold pencil-case, have found the contrary, to their cost.

There are many respectable men bookmakers—quite as good as many of the backers, in any respect; and if occasionally, for a few moments immediately preceding a race, the "ring" slightly resembles (in the present day) the Zoological Gardens at feeding-time, it must not be forgotten that it will not require an opera-glass to

distinguish amongst the crowd of anxious or hungry
candidates quite as many of the backing fraternity;
and that upon such occasions the bookmakers rather
represent the keepers or feeders. If, however, some
few of its members, who have been blessed with more
powerful lungs than others, could be prevailed upon,
when laying against ' Mario,' ' Tagliafico,' or any other
' O,' to modulate slightly their tones, whether the bass,
baritone, or tenor, à la Lablache, Santley, or Sims Reeves,
within moderate bounds, they would have a more pleasing effect upon the ears of the ladies, at least, " in the
dress-boxes," who would retire more gratified with their
entertainment, and with greater admiration of the racecourse. If a few, who in their anxiety to excel in
dexterity, and represent their acrobatic feats at the risk
of becoming impaled on the spikes of the enclosure,
think proper to do so, they should bear in mind that the
Sovereign does not permit her subjects to make away with
their own lives, even from the top of a telegraph-board or
a judge's box.

It may truly be said that the universe is a stage, all
men either actors or spectators; that Destiny composes
the piece, and Fortune distributes the parts. But shift
the scenes. The Turf is the compass, the Racehorse the
needle, and when the masks are off and the actors
appear in their true colours, the play is Gold. Talk of
the benefits of an Oxford or Cambridge education ! They
are wonderful as they are beneficial, if their objects be
alone to fit the students for domestic or idle life, or
more active pursuits, where there may be little or none
of worldly knowledge called in question. Invaluable to the
youth whose anxious parent may long to see him a Bishop,

or, perhaps, "a stickler for the Senate and the Forty." In the rudiments of human nature he will unquestionably become an adept by theory ; but his education will be no-where so highly finished as a perfect man of the world, than by a practical experience upon that stage, which truly represents in its genuine colours the principal object of life. There they run in all classes, in their true colours, and their real form. *They all try.*

Where is the member of the "honour and glory" division who would accept the first place in the race on the terms of handing over the "idol" to the second? This is the play in which all the actors display to the utmost their most brilliant talents, and vie with each other for excellence. It is their "benefit" wherein "each hero all his power displays." Therefore those who wish to study similar parts can best learn where and when the piece selected for the occasion is "The Universal Idol;" the after-piece, "The Farce." Many a man who has never dreamed, when at those universities, that he would, in after years, form a taste for turf pursuits, having done so sub-sequently, has regretted that his father had not afforded him an opportunity of learning the task he had before him until his experience had been too dearly bought, and then consoled himself with reflection on the pity he had not "started" with a knowledge of human nature in a practical point of view, and wished he were about to "start" again. Men generally learn how to get through the world about the period of closing their career, when they remain no longer slaves of prejudice, passion, or fame, but attend to the main point of the compass, "true as the dial to the sun."

Strip it of the disguise which hangs over it, the

sophistry which even its most infatuated followers may cast around it, and it can hardly be alleged that the turf is the exception to every pursuit in life, wherein the principal objects of most of its patrons are vanity and aggrandisement. Its most virtuous followers, in the winter of life, have become more perfect men of the world, having come in contact with not only the pure, but the depraved. They have had the best opportunity of forming opinions of friends and foes, as well as what is most difficult and frequently mistaken of all—of themselves; for there they are best tested in every form, in the thirst of man after those fashionable and worldly idols.

While it can boast amongst its patrons some of the highest, purest, and most noble-minded men on the face of the globe, there are others who carry their years well, their heads high, make a goodly show, and "keep their perpendicular." Still, while the bloom of youth has vanished from the cheek, the "brow of snow" hardly remains, however well clad in those fashionable attires which grace the form of the sycophant and adorn the perfect man of the world. The life of the zealous follower is a strange one, a continued scene of excitement and vicissitudes, without which life to him would be misery, and which in time becomes a sort of second nature. Yet it has, with many of nervous and excitable temperaments, between pleasure, anxiety, and pain, the effect of "tearing life out of them before their time."

The mind becomes almost completely occupied with reflections in some respect relating thereto. The adherent becomes infatuated with it, and its branches, which are many. How anxiously he awaits the publication of the weights for a great handicap, and casts his eye down the list in search of his own horse's name, which he hopes to find as near the bottom as possible; but should he find it sooner than his expectation, and contrast the weight with that of some others, which in his estimation are better in by seven pounds, or perhaps a stone, how suddenly he offers up a prayer (?) for the hapless handicapper! Should a thoughtless waiter, in laying breakfast upon a Saturday or Sunday morning, forget "Bell's Life," his pecuniary interests, as well as the bell, would be placed in jeopardy.

A curious circumstance happened many years ago at the country-seat of a gentleman, a zealous supporter of the turf. The Bishop chanced to pass through that part of his diocese. An extra Bible being required for the occasion, the clergyman sent, at an early hour on the Sunday morning, to request the loan of the family Bible. The servant (who could not read) handed a book to the messenger, who, however, returned with it in a short time, saying it was a "Racing Calendar;" the dialogue between the two resulting in the explanation, "that it was the book the master always read on Sunday."

Should an owner be fortunate enough to be possessed of good horses, he may rely upon plenty of friends—they will follow like a flock of sheep. "*Donec eris, felix multos numerabis amicos.*" One successful "tip" will make a dozen; one oversight or omission on a future occasion of a "good thing," lose fifty of the valuable (?) class (one hundred to be found any time at a London coffee-house); some of the number forthwith turn from the sunny to the shady side, and with a much greater *gusto* play their part in the farce than when favoured, which merely consists in the "Congratulate you, old fellow!" the slap on the back (the owner should be a man of nerve for his demonstrative friends). "What did you do?" merely casually intimating the amount of the "secondary consideration," which they may have themselves realised by the happy event. In the shady side, however, the case of the dog and his master is generally performed, where the former, being well fed and cared for during the month, was forgotten the thirty-first day, and bit the hand he had so frequently licked.

One of the seeds of trouble in such matters is that of *volunteering* opinions of the probable success of horses, some appearing to fancy that if an owner says he will win, "he is bound to keep his horse well and do so;" otherwise, if beaten a head, is at once put down as an "ass," or something worse.

The popularity, as well as the purse of proprietors, is subject to the turn of Fortune's wheel; the dame being, in that respect, extremely fickle. One day an owner is lauded to the skies, and his success hailed upon all sides—perhaps not more than merited—as a staunch and liberal supporter of the pastime; but should he chance in a hasty moment to adopt a course, probably practised from time immemorial, without comment by others, hey, presto! almost before the echo of the cheers which rent the skies have died away, the least touch of the wheel sends forth a thunder of abuse, and amongst the loudest blowers number some of his previous worshippers. Up starts another favourite, who is made an idol, until he may have transgressed to the displeasure of some of the populace, who must have somebody to vent their adoration upon, or to inflict with their loud expressions of

approbation, like the waves of the sea as described by the
poet,—

> " And one no sooner touched the shore and died
> Than a new follower arose."

Is it a matter of wonder that the animal, the instrument of all those extraordinary scenes, should have
conquered cruel Emperors, been made a Consul of,
and had cities built in his honour? And still he, so
frequently affording amusement, sometimes becomes the
apple of discord. But of how little concern is it to the
noble animal himself! He cannot change the world from
one of grumbling and discontent. *Nemo vivat contentus
sua sorte.*

Whatever may be the ups and downs, the pleasures
or disappointments, in the life of its followers, the
sportsman's motto should be " *Nil desperandum.*" He
invariably has his heart in the right place. " On one
fight more," said Robert Bruce (when almost driven by
reverses to despair), " hangs the independence of my
country;" and Bannockburn told its tale. Even the beginner may find himself the successful candidate for the
Derby, Oaks, and St. Leger, while his more experienced
opponents, after an age of disappointments, may end their
career like the monkey at Donnybrook fair, that so repeatedly jumped through three hoops without touching
any one of them.

But any forsaken child of fortune desirous to get a
peep at "Old Nick" on earth has but two courses to
adopt,—fly to the "rosy god," and then, under his auspices and powerful influences, proceed head-forward to
ruin. If ever there was one pursuit in life above any other
that required a cool head, and that it should be *properly*

I

*screwed on,* it is the turf. Lemonade or Seltzer water, cream or lemon ice, being especially recommended as a substitute for alcoholic beverages, particularly during the height of the fever, which so frequently accompanies its zealous patrons, who, in a pecuniary point of view, fall victims to the influences of over-excitement. A deviation from which recipe is sometimes followed by unpleasant dreams, which unfold a tale that the victim had, while a somnambulist, walked out of a prudent course, and had been led astray " by the invisible spirit, which steals away the brains."

To the turf are frequently attributed errors and misfortunes which never belonged to it. Many have been successful, both as owners as well as speculators ; many have not only had unbounded opportunities of embracing the " dame," when she even waited for them with extended arms, yet recklessly declined the proffered pleasure. Others have benefited by the opportunity, yet have cast them away upon worthless objects, and become the dupes of sycophants and sharpers. Others have been debarred from the possibility of doing so, Destiny having ordained otherwise; perhaps from the fact of the recruits having been over-confiding in persons, whose false fair face may have led them astray. They may have placed themselves in the hands or under the guidance of some old generals, perhaps grown aged in active service, whose doctrines or ideas of management may have consisted in the science of turning their weapons against the confiding owners for their own benefits, as well as made them the targets or outposts for others to fire at and practise upon. Still there have been instances where some of the most experienced old generals have, like

Stonewall Jackson, been shot with their own ammunition, and by some of their own brigade.

The turf abounds with not only opportunities for all, if properly embraced, but also with most upright and fair men in all classes, who in protecting their own interests, which is but natural and just, do not sacrifice every feeling of right and equity towards others, or permit mercenary motives to swallow up their entire thoughts,—men whose humane and kindly motto is " Live and let live," and who do not permit the hardened love of gold to render their hearts callous, their thoughts dead to every feeling beyond it. Yes, that pursuit, so well and justly termed "the glorious pastime," is stocked with the most noble-minded men under the sun, not alone noble by name but by nature, who really feel towards their fellow-men with Christian, and at least friendly feeling; and most of whom love in their hearts that noble animal, so justly prized by all true sportsmen, yet uncared for by others, who would not feel regret, further than for his price, if they saw him shot through the head before the sweat had dried upon him, or the noble blood in his veins had cooled, which he had just heated to the utmost in their behalf.

Reader, should you feel disposed to join the chace in the hunt after " the universal idol," take advice, and leave the " honour and glory " to take care of themselves. They *must* follow. Be not a " perfect man of the world," when the autumn or winter of your days has arrived, and your lamp of life flickering or about to be extinguished. Never start or sound your horn before saying your prayers. When you have a clear field make use of it, in order that you may not become whipper-in. Beware of the " brandy-flask," the sign and sound of the dice-box—the incurable cancer of the turf, the certain ruin of its followers. Let your cigar-case bear upon it the picture on Mr. Shandigaff's snuff-box. Avoid overlarge fences, or you may come to grief; and beware of riding amongst a crowd even of overkind friends, some of whom might be dangerous, although most experienced horsemen. When badly mounted, give him to your whip for a hack. Dispense with *managers.* When you are at fault read the remarks in these pages; read them through, and rest assured you will be more likely to *commence* with the "Blue Riband" and success, than *end* like the Donnybrook monkey and be laughed at. You will in after years, if you do not become a misanthrope or a sycophant, admit that the intrinsic value of half mankind is best estimated at the price of the *crêpe* on their hat-band, when mounted on the death of a rich relative, the wearing of which is about their most unsycophantic and least ungrateful act—that the world is a bubble, which soon bursts—and that the words written by that great and noble-minded man were thoroughly true,—

> " Glory, the Grape, Love, Gold—in these are sunk
> The hopes of all men, and of every nation."

# THE RACEHORSE.

Mr. Squeers' pupil merely enlightened his examiner, so far as related to the horse in general, when he replied that "*it* was a beast." Probably his answer would have been the same had his information been required upon the subject of a "donkey." Any remarks which I may venture to make will be confined to the animal so frequently called out of his name, "The Racehorse." The subject must be a simple one, inasmuch as so many attempt it; which is partly my reason for making a few observations and trespassing upon those parties who, in their leisure hours, may condescend to peruse them. Never having been overstocked with modesty, and having reserved sufficient assurance, perhaps, to fancy that about twenty-five years' experience may render it unnecessary to commence as lawyers generally do, when paid exorbitant fees, by an assurance to a judge and jury of their perfect conviction in their inability to do justice to the case, and their belief in the superiority of so many of their brethren, upon whom they so sincerely regret the task had not fallen. However, as my fee is but moderate, I shall consider myself as having given value in *length*, at least, if not in *depth;* and perhaps my best excuse will be to reply as the Irishman did, when asked why miles in his country were so long—"that they were not in good order, so they liked to give good measure."

There are, no doubt, many more competent than I to write upon the subject who do not, for various reasons, think proper to do so ; and there are others who do occasionally favour the public with most useful and sound opinions, evidently founded upon practical experience. Having devoted for many years almost my entire thoughts to the subject, I must confess to having failed in arriving at a conclusion upon a few points relating to the Racehorse. However simple at first thought they may appear, still not only have I failed, but the more I reflected upon the subject the more it puzzled ; leaving me to discover, as one endeavours to discover the perpetual motion, what so noble an animal was in reality intended for—an animal that can do almost anything but speak and discount a bill ; and why he bears so marked a contrast in every respect to that extraordinary animal referred to with the long ears, and frequently called Neddy, with which it has been ordained that he should breed ; yet that the produce of the alliance should totally fail, and be debarred thereby the possibility of increasing the similarity by further alliance with either side, although the strange and ominous cross upon the shoulders should distinctly remain to human view, but partially removed by the first alliance. There is one incontrovertible fact, however, that the racehorse has been instrumental in filling the purses of many, and generally of those who knew least about his real merits ; aye, has been the instrument of *whisking*, in the incredibly short space of a few hours' change of ownership, tatterdemalions from

previous obscurity into high position—in their own estimation at least—probably by a cup, or some other victory
of consequence; and thereby, in some sickening instances,
has given a sort of *locus standi* whereupon to ground an
amazing stock of bumptious assurance, even in the teeth
of the first nobles in the land. Probably such individuals
fancy, that to " get gold" a good deal of " brass" is occasionally necessary; while others, equally or more successful, bear their good fortune with becoming sense and
prudence.

In my remarks under the head of " Brood Mares"
I have stated that, from experience, I have found that
the time to commence to breed a racehorse, and insure
success, is to begin *" before he is born."* There lies *the
foundation,* and there *alone ;* the neglect of which rule is
the chief, if not the sole cause, of so many moderate, if not
useless, animals. As to *perfection,* it is out of the question
to expect it, if this course be disregarded; yet it is truly
incredible how many neglect it, and to what an extent;
sometimes almost approaching starvation. The dam is
the fountain, and, if neglected, the produce which she
may be carrying will not only show it, but *prove* it subsequently more plainly, at least to the eyes of some. Yet,
to those of any person who has made it his study, it will
be at once apparent when the foal is an hour old.

The opinions of many persons as to the care and attention necessary to breed and bring the racehorse to perfection fall very far short, indeed, of what it really requires.
An animal may look well; the dam or foal may *look well,*
but that is not sufficient. A foal may be larger or *taller,*
when dropped by a half-fed mare, than if she had been
properly fed and cared : but it will not have the condi-

tion, the lively spirits, or smooth, glossy skin, and healthy appearance, which the produce of a mare kept in tip-top condition will plainly show.

"For good or evil burning from its birth,
And like the soil beneath it will bring forth."

It should be remembered, that almost fabulous amounts are lost and won by the issue of events, which frequently turn upon, and are *decided merely by the nose* of the animal in question. How, therefore, can even a breath of air in climate or temperature, a feed of corn, or a particle of attention, be economically and prudently disregarded, especially in these days of competition for excellence?

Has the reader ever asked himself, or reflected upon the question, as to what kind of animal the first horse was? Was he in the form of a racehorse, a waggonhorse, or a hunter? If not a racehorse, how has he become one? who has brought him to his present formation? and by what means? The Creator formed the horse, as He did man, in certain shapes; man has worked out of the materials the remainder. Take two own brothers or sisters, in every respect as nearly alike as possible; and if the object be to send each in different directions, to preserve the one in its present perfection and reduce the other by degrees, and make them dissimilar, he has only to treat them differently as to care, feeding, temperature, and exercise. Each could be thus rendered, by time, as dissimilar as any two animals of the same species could be. Look at Shetland ponies. What are they? Where did they come from? What reduced them in size? Take the powerful float-horse, well fed and cared; he has been forced in growth and power; still his limbs do not possess that elasticity, nor have they been refined down, lengthened, or rendered active by training, as the racehorse. There are many thoroughbred stallions that possess equal, if not more power, to many of these draught-horses; although their bone and muscles may not appear so large, yet they are of better material.

There never have been such splendid specimens of the thoroughbred horse as are to be seen in the present day, the descendants of those two extraordinary mares, ' Guiccioli' and ' Pocahontas,' from the ' Whalebone' or ' Sir Hercules' strain and ' Bob Booty' blood; for it cannot be denied that ' Irish Birdcatcher' has done more for the racehorse than any stallion of modern days — probably than ever was heard of; not alone in speed, but in symmetry of shape and power. The ' Bob Booty' mixture has told the tale. In symmetry the ' Sweetmeats' alone rival the ' Birds.'

I have been from childhood (the days of blowing soap-and-water balloons from a long pipe) amongst thoroughbred horses, and, during my schoolboy days, for years in the habit of weekly visiting at the residence of a friend of my father, who had not less than one hundred thoroughbred horses, including brood mares, foals, and animals of all ages — some of the best blood in the world, where I had been many a time found watching and amusing myself with them. An extraordinary circumstance then and there occurred, during one of my visits, which proves the uncertainty of young stock, and how frequently prejudiced and hasty opinions are formed with regard to them. The gentleman referred to now resides on his own estates, as he did during his father's lifetime, who was

also owner of many first-class and celebrated horses. Both
were first-rate judges; and if ever there were two owners
who really kept horses for honour and glory, and really
were fond of them in a sporting point of view, they were.
At the time referred to, upwards of twenty years ago, a
certain renowned horse upon the turf, in both countries,
had been put to stud; the attention of the sporting million was turned towards his yearlings; people from various quarters came to see them. My friend happened to
have some of his stock, and amongst them a most promising colt, the produce of a celebrated mare. The
owner of the sire was so proud of this colt, that he offered
to back him against any other in the world, for any reasonable amount; numbers were asking permission to see
him, and large sums had been refused for him. In a
back-yard, however, far removed from observation, and
comparatively from attention, there happened to be a
certain yearling by an unfashionable sire, of rather
plainish although good shapes; but he had one friend
about the establishment, who used to bring, with his own
hand, many a good feed of corn, viz. the writer. The
owner, finding me continually in the stable, and invariably
a manger-full of corn before the colt, one day laughingly
inquired why I was so fond of him. " Why," I replied,
" he is worth a dozen of your crack." Whereupon he
asked me if I would purchase him, and offered to take
thirty pounds for him. The yearlings were all sent into
training; this one, after some consideration, amongst the
number. The crack turned out perfectly useless for anything, although sound; the other, one of the best horses
that ever trod the turf, which he was considered by many
of the best judges, who stated their belief that he could

have won "nine Derbies out of ten," and of which 'Chan-ticleer,' 'Eryx,' 'Cawrouche,' and many others, bear testi-mony. He was named after a favourite wine of his owner. It may be a sort of encouragement to beginners to know, that the best horse I ever owned was the first thorough-bred yearling I purchased, and that his price was seventy pounds—'Bright Star,' one of 'Irish Birdcatcher's' first and best sons. He shone brilliantly only in the Emerald Isle, his engagements being confined to that country, where he extinguished the light of all of his own year, including 'Peep-o'-day Boy,' and horses of all ages.

Probably one of the most important subjects in con-nexion with the racehorse is the mode of purchasing and selling, a subject surrounded with many important con-siderations and difficulties. There are no dealings which lead to more disagreements in various ways, or so fre-quently cause trouble and misunderstanding.

The differences of opinion, as to value of young horses especially, are so wide, that it is unnecessary to dwell fur-ther on that point than to state my opinion that there is as much difference between some sellers or markets, as there is between dealing in a shop in the Burlington Arcade and a back street in London. And as to the prices, as great as between hotel charges to a racing-man and a commer-cial traveller: a great deal depends upon the fashion and name. (One of the greatest public dancers for many years, who had danced to the top of the tree of fame on the stage, or whatever the pitch of perfection may be among such *artists*, was really named O'S——n, an Emeralder. Nobody then thought he could dance a step. He took a certain Italian name, a most melodiously sounding one, and wore gold earrings; the consequence was, the houses were

crammed almost to suffocation; and during a conversation
he remarked, laughing, "Why, if I were not a foreigner
I could not dance." I have seen him myself frequently.)
Still, many first-rate articles are often picked up at
very humble establishments, and very great counterfeits
sold at some fashionable or flash ones, where the measure
of the purchaser's purse, as well as himself, is sometimes taken into consideration. Moreover, a great deal
depends upon the *humour* and the *time* in which the parties may transact business. A certain well-known breeder
and owner of racehorses, and of first-rate ones, now deceased, was remarkable for his peculiarities in such transactions. Although wealthy, and otherwise independent
in his views upon the subject, he was fond of selling. It
was well known, that the time to approach him was just
after defeat, when he fell amazingly both in spirits and in
price. His principal rule, however, was never to vary a
penny, or reduce when once he fixed the price: no one
who knew him ever attempted to offer less. "Yes or
no! and if you take him, you take him as he is; the moment he leaves this yard, he is yours. Examine him, if
you like; I'll give no opinion," &c. If a stranger offered
less he merely smiled, as much as to say, "You don't
know *me*." His other peculiarity was, that when he *did*
sell, he invariably had *better* in his stable; so much so that
the knowing ones always "pricked up their ears" when
they heard of a sale. Upon one occasion a friend of
mine, having some horses engaged in several rather heavy
stakes, and knowing that this gentleman had a certain
horse that could beat him, he purchased him for more
than a thousand guineas, and booked him a cheap horse:
he was, no doubt, a good one. Upon coming to the post,

however, within a fortnight afterwards, to walk over as he thought, he not only found his recent companion there, but had to lower his colours to him for all the engagements, amounting to more than the price of the horse. His victor, however, was one of the greatest horses of modern days in England, both as a racehorse and a sire. The fact is, as far as judgment of the animal and knowledge otherwise in purchasing, some men would do more with one thousand than others would with ten thousand pounds. It is astonishing the few even fair horses some persons get, compared with the heavy prices they pay.

Truly may it be said that bought experience is best. Perhaps the reader may fancy that the subject of the negociation in horse transactions is a secondary consideration. Perhaps so; I have found it a hotbed of annoyance, disappointment, and loss; all of which, *every particle* of it, has been brought upon me by placing too much confidence, and omitting to have every word of contracts distinctly and clearly written down, signed, and, if necessary, witnessed. Perhaps the following plain and unvarnished statement of one case, resulting from omission of the "black and white," would convince the reader of the necessity of adopting precautionary measures.

I once owned a two-years-old, a well-known horse, and a very good one. (I found him totally neglected, after being weaned, at a farm remote from the breeder's residence, and purchased him for a mere trifle.) After he had run and won his first engagement in Ireland, in June, I chanced to be at Liverpool races the following month, where several parties sought to purchase him, as he was engaged in the Derby. At the request of one party I sold him for 1000 guineas, repeatedly adding the words,

"Mind, the price is 1000 guineas, and I shall give no warranty further than 'wind and sight,'" although the horse was as sound as any animal living; adding, "that I should require the money cash down, and that before I sent over to Ireland for the horse, or concluded the bargain, I should speak to a certain nobleman who happened to be on the ground, with whom I had matched him against a filly of his for 500 sovereigns a-side, to be run in the following October meeting in Ireland. All was decided and agreed upon; his Lordship acquiesced under the circumstances of my having sold—as he remarked, "not to prevent the sale." I gave him fifty pounds forfeit to be off, and sent for the horse by a special messenger. The animal having arrived, and after refusing several parties permission to see him, who were anxious to purchase, I found myself in his stable with the purchaser, who did not profess to be a judge, accompanied by his brains-carrier, who considered himself a *nonpareil* in such matters. He, however, declared that the horse had bad fore-legs, whereupon I challenged him to produce any veterinary surgeon in England to test them. However, after some useless *badinage*, I requested the individuals to vacate the stable; directed the door to be locked, and declined further negociation at any price; which the parties subsequently broached with wonderful assurances of contingencies, with reduced price. I sent the horse his journey home, upon which he encountered a gale at sea, which terminated in the smashing of the horse-van, and as my man informed me, he would have taken ten pounds for his chance at one time. He got a heavy distemper, under which he suffered for months. So much for sale number one. A second party, an individual named Dr.

P——r, of Rugeley, Staffordshire, then sought to become purchaser at 2000 guineas, by letter and three telegrams, now in my possession, inviting me to spend a few days at Rugeley. The third happened to be made by a gentleman in high position, who, through a friend, became purchaser at 2000 guineas, having sent over a most experienced and respectable party to see the horse; and who, at my trainer's request, even rode him a gallop, and highly approved of him.

In consequence of receiving a letter that all was right, and that he would take the horse, I sent him to a most careful trainer's establishment, informing him of the sale. There the horse remained for a considerable time — about six weeks or so — during which time some of the stakes in which he was engaged, and of considerable value, were won by a horse in the same stable. Of course I could not run the other, having sold him; nor was he quite fit, not having recovered his stormy passage. Next comes a letter informing me that the gentleman had lost so much money at Newmarket that he could not take the horse, and requesting that I would look out for another purchaser. Being thus left upon my hands, and being engaged in a valuable stake the next month, and having to pay all the stake in any case, I started him: he won the race, giving ten pounds to the winner of the Doncaster St. Leger the following year, and a field of horses (including the filly against which he had been matched for five hundred a-side, at even weights, beaten a distance), although he was not in training, any odds against him, and not backed for a penny-piece. I subsequently sold half the horse; sent him in February to a most respectable and first-class trainer in the north of England, who was sadly disappointed at his wretched condition, owing to his illness during the winter. He made wonderful improvements under the circumstances; when the trainer on the eve of the Derby stated in writing his opinion, which is second to none, that no horse living, if ever there was one foaled, could beat him for the Derby. He was placed, beat the first favourite (at two to one), and was beaten about a length; the trainer remarking to me, that had he had him a little longer, and not been obliged to give him carrots to get him into condition before training, the result would have been different. I subsequently sold him : he won many races, beating the best horses in England, and ended his career by breaking his leg at exercise in five years afterwards, having struck it with his hind foot. The day he did so he was as clean and fresh on those legs, and as sound otherwise, as the day he was foaled. Thus ended the career of a horse that put many thousands in the pockets of others, little in mine.

Another great point to which any owner of blood stock, especially the breeder, should direct his especial care, is as to the parties to whom he may sell. It would be better, in some cases, to sell for half the money to some than to others for three times the price. Many horses being turned to such wholesale plundering purposes retire from racing with characters of impostors, and thus injure the sale of others of their breed to an enormous extent, their real merits not being half, if at all, developed.

In selecting young stock the novice would do well to observe the following general rules, with regard to shape, size, &c., setting aside the question of breeding. In the first place, size and substance are indispensable—not a great, tall, narrow "clothes-horse;" on the contrary, having more the appearance of being thick-set than otherwise (that is when in high condition, and previous to going into training), for it is astonishing how they fine down and lengthen after training and time : whereas the class before described, without such shape and substance, upon getting the necessary work, become "perfect shells," weak and useless, growing tall but not thickening. At the same time coarseness should be avoided, especially as regards the head, neck, and shoulders. The coarseness of the head generally consists in thick and ill-shaped jaw-bones, almost as broad at bottom, towards the nose, as at top, with a fleshy thickness; the eye small and sullen, which should be large, clear, and bright, with a sort of comparative baldness, or absence of coarse hairs around, which is in all animals, as well as the horse, indicative of high breeding. The jaw-bones should be shaped, tapering gradually towards the nose, clean, and free from superfluous fleshy substance ; a good space between the jaw-bones; forehead wide and flat between the eyes. (Here there are exceptions, many first class horses being the contrary.) The heads, in many horses of the first class, bear striking contrasts : some being plain and sensible-looking, of a clean, bony kind, not over-small or ponyish, with a clear, full, and steady eye, which generally denotes good temper, and staying and enduring qualities; whereas the fiery or anxious eye, which displays the white more than usual, is generally found in the flighty-tempered, speedy, but non-stayers; and more frequently with mares than horses. The ears are not so much a matter of consequence, at least for racing purposes, provided they are not of that long, upright form, like a donkey,

K

and stuck up at each side of the head. Some of the best horses
have had lopped ears, in some cases coming down over
their eyes like a rabbit; such as 'King of Oude,' 'Sir Tatton
Sykes,' 'Oulston,' 'Camobie,' and others.

I have always remarked that horses with lopped ears are
invariably very good-tempered, and good in other respects;
and it is equally true that horses with any peculiar fashion
or habit, such as hanging out the tongue at one side, over
the bridle, or rocking while in the stable, resting one hind
hoof upon the other, are generally good animals.    As
to the 'King of Oude,' probably a more extraordinary
example of curious shapes and formation never was foaled;
for although he possessed, when " dissected," many
capital points, and when looked over was a very fine outline of the racehorse, yet upon first appearance he gave
one an idea of being some species of animal never
before discovered.  He had a pair of horns, about two
inches long and an inch wide, and his ears hung down
about a foot long, as if they had been stitched on by a
cobbler for some particular occasion.  Taking him for
all in all, I fancy we shall never look upon his like again.

While upon the subject of ears, I have frequently observed that good and true animals, when cantering or galloping, prick backward and forward alternately the one or
the other; and I believe it to be a sign, that they are
happy and contented: and, moreover, they are invariably
long runners, and good in every respect.  The nostrils
should be full and roomy; reasonable length of neck, which
should be muscular and strong, but not coarse.  A very
short neck is generally accompanied by round, heavy, or
misplaced shoulders, as well as shortness in other respects,
which is the worst failing in a racehorse; for length is of

all things desirable. Speaking of length, a mistaken notion sometimes prevails in the minds of persons that it means a long back; whereas it is nothing of the kind. We must judge of length by the ground which an animal covers *underneath*, the placing of the shoulders, which should be well placed back in an oblique form, together with good length from hip to end of the haunch-bone, supplying the length where it should really be found. A slight drooping towards the tail is preferable to too level an appearance. Animals with such drooping shape are generally better turned under on their haunches, and possess more propelling power. Length to a certain extent, in every point, is necessary in the racehorse. The arms should be muscular, and reasonably long; but it is most desirable that from knees to fetlock they should be shortish, clean, and with good bone and sinew, not round and "gummy." The fetlock-joints, to insure "long standing," should be of reasonable length and substance; not straight or upright; for horses thus formed seldom last long, and are generally strait and tied in the shoulders and elbows—fatal points in the racehorse. If there is any inclination in either respect, arched knees, appearing rather bent over, are preferable to "calf knees," which have the contrary appearance, and cause generally an extra pressure on the back tendons. The body or middle-piece of the true-made weight-carrying racehorse, when in condition, should present plenty of depth of girth, good back, muscular arched loins; but the back ribs do not always present that power or depth, nor should they be coupled up towards the hip so closely as some persons appear to fancy: for the longest runners and best weight-carriers, as well as the most speedy, present

the appearance of being light in their back ribs. 'Alice
Hawthorne,' for instance, was very light in this respect;
perhaps more so than any other animal that could be
named. Still she had wonderful depth of girth and fore-
rib, which partly caused the back ribs to appear so very
light or shallow. But the great point of all for propelling power lies in the hind-quarters — good length
from hip to hock, the quarters being well placed under
good thighs and hocks. Where horses are very close,
and well ribbed-up, that is to say, where there is but
small space between the back-rib and hip, *the latter
being sometimes deep and round*, there is not freedom of
action, propelling power, and fine stride, which are found
in animals shaped as 'Alice Hawthorne,' and others of
her mould; and action carries weight. 'Thormanby' (her
son) takes after her very much in his hind-quarters, in his
great length from hip to hock, and his fine lengthy stride;
and her son ' Oulston' to a great extent does so, especially
in his fine style and freedom of galloping.

There can be no question that the formation of the racehorse, to insure action and success, must principally depend upon the propelling power; for, like the connecting-
rod in a railway engine, on that the machinery depends.
A level, evenly-proportioned, fair-sized horse, about fifteen
two, three, or say sixteen hands, is the style. Horses are
seldom really so tall as in their owners' eyes, the deduction
of an inch from his standard being about the true mark.
Long, low, and level is the best line to be guided by.
Perhaps one of the most important points to have regard to is the formation of the chest; for where you
find *the chest very broad*, and the animal "*standing very
wide on the ground*," you may pass him as "no race-

horse." On the other hand, some of the speediest, stoutest, and longest runners, as well as best weight-carriers, have been very narrow *between* the fore-arms and fore-legs; still they may have a well-expanded chest. 'Harkaway,' 'Alice Hawthorne,' and many others, were narrow in this respect; and 'Fisherman' stood and walked wide on the ground, still was not so above. Horses that are so heavily formed in the respect referred to are generally badly formed in the shoulders and elbows. At Doncaster, some years ago, I happened to be looking at a yearling filly, a daughter of a renowned mare on the turf and at stud, and since the dam of a Derby winner. Being alone for some time in the stable, remarking the peculiar breadth of her chest and between her fore-legs as mentioned, and turning in my mind the improbability of her proving a racehorse, a certain well-known northern trainer entered, accompanied by a friend, and at once made the same remark, taking no further notice of her. The filly turned out useless. Many similar cases have been seen. One being in that of a yearling, bred by myself; a fine colt of a running family, and well-shaped in other respects, and yet in a moment condemned by the late Mr. Watts on that ground; his remark simply being, " he will never be worth a bowl of soup."

Many animals are flat-ribbed, and have an extraordinary appearance of weakness behind the saddle, being of great length in that respect, yet many of them stay well. Some of the 'Touchstones' are remarkably so : but they have most muscular loins, fine quarters, and thighs, with propelling power; their hind-quarters well turned under, and they seldom throw curbs. The only instance I can remember was one in which I was the sufferer; which will go far to prove to the reader the necessity of making inquiry, and, as far as possible, learning the antecedents of any animal whose produce he may be about to purchase.

Amongst a number of yearlings for sale at Doncaster there chanced to be one whose blood was undeniable, and during my rounds of those advertised for sale, handbill in hand, I immediately marked one from a knowledge of the dam and sire. The colt was by 'Touchstone,' but was not exposed for public inspection the evening preceding the sale, although the rest were.

Just as the sale was progressing next morning I chanced to be approaching the ring, and saw a fine colt being led round within the crowd, when a party, who had for years haunted me like a bird of ill omen, whenever he wanted a favour or had bad news, came rushing to me and in most laudable terms exclaimed, "Don't lose him at any price." The hammer was about to fall the last time. I made an advanced bid in a hasty moment, and upon entering the ring at the first look beheld a colt dead lame, knuckling, and hardly able to walk, with ringbones, curbs, and spavins, and spent an anxious moment in suspense (hoping the owner, who was near me, or some blind speculator, might advance and permit me to retreat), when down fell the hammer, resounding the sale to me of the highest-priced colt (double the price) I ever purchased, with his engagements in Derby, St. Leger, &c. To crown all, having sent him to a celebrated V. S. in London to practise upon, he informed me he was broken-winded. I sold him for thirty shillings. The *Connoisseur* previous to the St. Leger, while I was blowing him up on the subject, expressed his surprise that I was losing my time instead of being engaged laying against "the lamest horse he ever saw walk or gallop;" that he had seen him at exercise. The horse won the St. Leger "in a common canter." The following year a friend of mine requested I would accompany him to see two yearlings he had purchased. I did so. The one was out of the *same dam*, good-looking in many respects, by a first-class sire, but with some sad deficiencies where perfection was most necessary. He paid a high price for her (some hundreds), and upon being informed of my misfortune, and that his was one of the breed, he did not seem gratified at his bargain. This lot turned out useless. The following year another colt, out of the *same mare*, was put to auction, and bought by an intimate friend of the last purchaser for a *very high* price indeed, but also turned out good for nothing. Since then I have from curiosity watched the sale of her produce, and they have absolutely continued to fetch high prices; while others, really valuable, if they were forced upon people might not meet with a purchaser;— like 'Thormanby,' who was exposed for sale all the week without avail, until picked up for 350*l*. through the sound judgment of his experienced trainer. 'Voltigeur,' and many others likewise, in the same way.

Upon the subject of purchasing, it is a strange fact, that the fall of the hammer within two minutes has placed to the credit of the auctioneer, by the sale of that noble animal, the racehorse, a sum exceeding the salary of some of the younger but poorer branches of the best families in Great Britain, in Government Offices, where they toil from New-year's Day to Christmas Eve, and which appointments they obtain through Ministerial influence, provided they pass a strict examination as educated gentlemen.

One of the most important points to which the attention of purchasers should be directed is "temper;" nothing being more hereditary or more fatal to the racehorse than the want of it. It is therefore most necessary, as far as possible, to arrive at information as to the antecedents of brood mares in this respect.

"Birds breed not vipers, tigers nurse not lambs."

Still a predisposition to bad temper may be considerably overcome, if not completely eradicated, by proper treatment; always taking care to place young horses under the charge of steady persons, who will not, on the one hand, play tricks with them, or, on the other, abuse them : for want of heart or pluck in either man or beast is fatal, in many instances. They should not be allowed to master on the one hand, and *vice versâ*. A proper medium between kindness and chastisement should be observed. There is nothing like shaping the sapling when young. Some young as well as old horses, particularly fillies, naturally if not absolutely bad-tempered, still are faint-hearted and soft; and although the colts in a family may be quite the reverse, the fillies are frequently objectionable in this respect.

I once purchased a yearling filly, sister to 'Wolf-dog' (a horse of superior merit, once the vanquisher of the celebrated 'Alice Hawthorne'); she not only gave every promise of turning out a first-class animal, but, in reality, could run in private. At the commencement of her training I happened to meet a party (who trained during many years for the breeder). Upon speaking of this filly he jokingly remarked, that she was good-looking and well-bred enough for anything, but recommended that she should be kept in an outer yard, as far away from the house as possible; for, added he, "if the cook should happen to come outside and blow her nose, *divil* a feed of oats she'll eat for a month." His estimate of her qualifications and imperfections proved perfectly correct, for, like several other fillies of her breed, she was a nervous, soft-hearted jade, although a beautiful animal to look at.

A certain well-known breeder and owner of racehorses for many years sold his fillies for "mere songs," as a rule; such was his dislike to them. Young stock are frequently kept too much confined, and do not get half the exercise which they should have from the very day they are foaled; for, bearing in mind that, like other animals by nature formed for speed, it is natural to suppose that their muscles and joints must be properly afforded the means of development, and every facility to promote freedom of action: in fact, to "*run faster and show more speed than any other.*" Suppose, then, that deer, greyhounds, hares, foxes, or any other animal by nature speedy, were to be kept confined and denied liberty to exercise, what chance would they have, reared in that domesticated manner, with those in their wild and natural state? The foal, from the *moment it is dropped*, should have plenty of space to give freedom to its limbs, and avoid the possibility of its becoming in the slightest degree confined, or its freedom of action lessened. Confinement, or want of expansion of the muscles and joints, must materially deprive the animal of its natural speed; for common reason dictates that the racehorse, above all other animals, must not and cannot be an exception. My particular reason for directing the reader's attention to this fact is, that as everybody knows, or ought to know, that "practice makes perfect," how can people expect their "caged-up," half-exercised young ones, to equal or contend with those that have, perhaps, not only had their natural comforts attended to otherwise, but in this respect their paddock to exercise in unmolested? not allowing, on the one hand, their joints to become stiff from want of ordinary exercise, or, on the other, to become strained, tending to cause curbs, spavins, &c.: which might be the case if left with other young ones, for they frequently hunt and run down each other if kept together, and injure their wind through excessive galloping. The exercise should be given without overdrawing, spraining, or otherwise impeding progress in any respect. Such omissions frequently happen, yet the consequences and losses do not always fall upon those who are in fault, as the effects of such neglect do not at all times appear suddenly; on the contrary, perhaps in a week, month, or considerable time, after the animals have changed hands.

Bearing in mind that it is much easier to become a purchaser than to find one, parties should be cautious before investing; which is not attended alone with the first outlay, but if young ones are good-looking, and well-bred, they are generally heavily engaged. Many are made up for sale, and look well to the eye, being apparently fine, grown and well-fed, yet in reality have lost to a great extent their natural freedom of action, from the fact that they have been forced to look well to the eye, yet are not in really good condition or fit to go into training; some being so fat inwardly, that they often go amiss in various ways upon getting proper exercise or being trained, and sometimes become affected in the wind. Overfed and housed young ones are not the most desirable.

Another benefit resulting from the inspection of the dam and as many of the family as possible is this, that many yearlings are handsome and well-formed, when, quite suddenly, they take a change to an amazing extent, sometimes improving with rapidity in faulty respects, and at others growing quite the contrary, until finally they become perfectly metamorphosed. Just like mankind: we frequently see children with regular Paddy-noses; and yet, when they grow up, they become perfect Wellingtonians. It is, therefore, most desirable for a purchaser to have a knowledge of the sires, dams, and as many of the family as possible, when purchasing yearlings, because they change so quickly and to such an extent: moreover, an experienced breeder, in crossing mares with sires, seeks in every respect to amend deficiencies, whether as to temper, size, colour, or shape, &c.; which improvements may not be fully developed at a year old, but still are most likely to appear in time.

> " The tainted branches of the tree,
> If lopped with care, a strength may give,
> By which the rest shall bloom and live,
> All greenly fresh and wildly free."

Many yearlings are undersized, yet grow wonderfully, if the sires and dams are large; whereas, if *forced* for sale, some become afterwards what may be termed *stunted* in growth. They may *thicken*, but they will not *grow* in proportion. The latter frequently happens where they receive a temporary check in growth from having changed into bad hands, or from sickness, or want of being kept in *really good health:* for animals may, by forcing or artificial feeding, grow in every respect; still they may not be in perfect health: and even brood mares may feed better than others, and be in bad health, through inward complaints, and die suddenly therefrom. I have seen balls as large as cannon-balls, and almost as heavy, that have been taken from them on *post-mortem* examination, and preserved, polished, and cut in two; when they represented beautiful pieces of grained marble, which was composed of sand, hay, straw, &c. These curiosities are to be seen at some veterinary establishments, and are well worth looking at. Such is generally the consequence of not cleaning food properly, or want of occasional medicine; and those balls have been forming for a long time. The very best, most simple, and least dangerous medicine, and one that I found valuable beyond description, is linseed oil, especially for yearlings, where their coats are found cold or staring, or the animal not putting up condition; and for all horses or any animals there is no food more beneficial than a little linseed meal or oil-cake, for putting up condition, and otherwise improving health. But with yearlings especially I found the *oil* make *immediate* and *marked* improvement. The coats that had been staring, in an incredibly short time became like satin, and the animals in all other respects made vast progress.

If the dam is unhealthy she cannot afford the nourishment, and what she does yield will not be so good; yet half the owners of mares overlook this important point of absolute health. How frequently do we see mares "mowing down" capital pasture, yet look cold, and not put-up condition! It is quite true that many dams, the best nurses, will yield so much to their produce that they lessen their own condition thereby; but still there are many bad nurses, as poor-looking as if they were half-starved, and especially staring and cold in their coats, and " hide-bound," like the bark on trees—the certain proof of ill-health. Still, many owners say, " Oh! she is all right; she has plenty to eat."

Temperature and climate are most essential to *perfection*; every breath of air and change of wind are of consequence. And here too much care cannot be taken in changes of weather. Sometimes the sun may be shining, suddenly cold winds set in, and here is where the attention is required, which proves how frequently men have a greater number of such animals to attend to than they can possibly do justice to: that is, if *perfection* be the object. To my mind, it frequently happens that there are about half a sufficient number of attendants upon such stock; whereas I know instances quite the contrary, and as much difference as between chalk and cheese in the results in every respect, when compared with those of the "do-well-enough idea." It is subsequently they *prove* it, when it comes to the struggle to a *nose*.

It appears to astonish some people that the French should defeat the British in horseflesh. It would be a source of amazement to me if they did not. If asked why, my simple answer would be in the Irish fashion, by asking another question—Why should they not? Tell me one single point in which they do not come quite up to, if not absolutely excel, the British in every thing necessary towards success, as far as the animal is concerned, up to the time he leaves the starting-post? Then, as to their talents otherwise in the management—their trainers, their jockeys, their judgment, and their knowing how to take care of the " main point" (in which they are quite as 'cute as the Yankee),—in my humble opinion they can give them weight.

Their selection of trainers and jockeys — who are English, no doubt — do them credit; and their climate speaks for itself. As to the fallacy of saying, " Oh, they are English horses," it is moonshine ; they are nothing of the kind. One might as well say, " Oh, they are not English horses, they are Irish." Pray, where did ' Faugh-a-Ballagh' (sire of ' Fille de l'Air') come from? ' The Baron,' grandsire of all the best horses in England, from whose sons ' Stockwell ' and ' Rataplan' have sprung ? the grandest specimens of the racehorse ever beheld in every respect ; besides many in former years that never were heard of in England ? If I mistake not, steam will carry an owner or a horse in as short a space of time from luxurious London to princely Paris as to desolate Dublin, and with half the sea. If there be superiority, no matter how or why it exists — there is no use mincing the matter — the French have it, and the point turns on " climate." The reason they have the opportunity of availing themselves of that benefit is simply open markets, free and willing purses, and good buyers. It would, indeed, be strange if the land which is sought by the declining Britisher, to restore, by the soothing influences of its climate, the declining constitution, and which yields for his consumption the delicious and nutritious grape, could not, in this respect, vie with "the tight little island" so celebrated for beef, barley, and juniper juice. Yes, and as certain as those remarks are penned, there will be more demonstrative proofs of the fact. It is an old adage, and a true one, "that every man knows best where his boot pinches." I never in my experience received such a shock in horse-racing as I did through a " Frenchman," who appeared to come like a thunderbolt from the skies; and he was not only a 'Prétendant' but a 'Faugh-a-Ballagh,' for he "shot" me by a "head" with his blue cap on; which was all I could see, and did not seek for more, deeming him anything but dangerous, and being more afraid of "Gallus." With regard to the allowance in weight, instead of being looked upon as a compliment, it should rather be taken as an insult, or a piece of vanity; like a school-boy telling his playfellow that he would tie one hand behind his back and then thrash him. Let it also be borne in mind, that the Emperor is not only an en-lightened judge of horses, but a staunch supporter of the turf.

With regard to a horse's colour, a great diversity of opinion exists. It has been said, that "a good horse never was of a bad colour." Still, it will hardly be questioned, that there have been more good of certain colours than of others: then, assuming that particular colours have a tendency towards enabling a purchaser to become possessed of a good animal, the subject must merit attention. We must arrive, therefore, at conclusions derived from practical experience; and admitting the fact, that some are more general or common than others, let us first take those of gray and black, which are the rarest, especially the latter. Where can we find (as far as mares are concerned) instances in modern times of good animals really black, with a solitary excep-tion or two, such as 'Priestess' (the 'Doctor' blood)? Curiously enough, black mares usually, and more fre-quently those of a very dark brown, almost black, are of the 'Touchstone' or 'Sir Hercules' blood. The very best brood mares being of that very dark brown colour, with a mixture of gray hairs in the flank and tail, which denote generally the descendants of the 'Whalebone' blood. Although 'Saunterer,' 'Nunnykirk,' 'Black Tommy,' 'Launcelot' (brother to 'Touchstone'), and others were good, we seldom find good mares for racing purposes. It is extraordinary, considering the number of the 'Whalebone' blood now bred, that so very few are black; and yet the generality of those black thorough-bred animals are descended from that blood, so remark-able for stoutness: still, animals of this colour are inva-riably soft, mares especially.* Although we occasionally find a few good gray horses, yet, comparatively speaking, they are, "like angels' visits, few and far between." As to the merits of bay, brown, and chestnut, each have afforded so many proofs of their excellence that it would almost "puzzle Paris" to whom he should award the golden apple. As far as the writer can form an esti-mate, the Derby has been won within the last thirty years by seven chestnuts, seven browns, and sixteen bays; the St. Leger by five chestnuts, eight browns, and seventeen bays; with about the same proportion in the Oaks.

Many a valuable young one has been sacrificed through want of that attention which is so needed, par-ticularly in the case of the racehorse: for in point of fact, according to the present state of things, one cannot be too particular in the respects referred to. An erroneous idea exists in the minds of the proprietors of many establishments that one man can attend to a large num-

---

* For my part, I am not so very partial to, nor a great believer in, the staying qualities of many 'Touchstones;' on the contrary, believe that their *forte* is speed, and their best point that they are of a very running strain.

ber of animals : if they can, it depends upon "the sort of minding."

One of the most important, if not *the* most important rule to be observed, is strict attention to the feet, no drawback or neglect proving more injurious; for if the power or gift to run faster than any other animal be the object, it is easy to imagine the sensitive feelings, especially when galloping upon hard ground, when they have been neglected. In the races for the first and second classes of the Madrid stakes at the Curragh April meeting, when 'The Baron' (sire of 'Stockwell') and 'Highwayman,' ran with many others, the result was that the latter won the *first class* in a canter. I shall never forget the veteran Mr. Watts' astonishment at his horse's defeat as he leant on his large twisted stick, believing him invincible. In two days afterwards, for the *second class,* same distance, &c., 'The Baron,' reversing their positions, won easily; the owner having just previously backed him for a ten-pound note (his favourite investment) for the second journey; and seeming quite pleased, explained to the bystanders that he could not account for the horse's running on the first day, but had him plated for the second occasion *with a piece of thin leather between the plate and hoof* (to the want of which he ascribed his horse's defeat on the previous occasion), having had thin, shelly soles, or tender feet. What does it matter how good horses may really be by nature if, through neglect, accident, or ignorance upon the part of the owner or care-taker, they are lost or become useless? Many horses are parted with as worthless, from ignorance of the real ailment or causes of their indifferent performances, which may proceed from the most simple reasons

L

possible; and what can tend more to retard speed or action than anything wrong with the feet? Bad shoeing in many instances being the cause.

The very time above all others, when the experienced hand, as regards the trainer, is called in question, is when the animal is commencing or being taught "the way he should go." The foundation before described, as to care and condition, should be laid, in order to leave something to work upon; the first attempt of the trainer being to get rid of the "soft foal's flesh," and replace it with as much muscle as possible, thereby developing, as far as the age of the animal admits, the natural shapes and powers, without reducing the frame too much or too suddenly. Especial care should be taken not to draw the young ones too fine, for they run better when as "big" as circumstances will admit; and, naturally, will not bear to be reduced in condition as old horses. A mistake sometimes made is the supposition that, because a two-year-old feeds well, and keeps up condition, he can stand extra work in proportion. To a certain extent they, like all other ages, must get work according to these rules; but a two-year-old may be a fine-constitutioned animal, inclined to carry flesh, and what is termed "a gross horse," and still, if worked and galloped according to the general rule, will prove that, although he may feed well and thrive upon work, still, if the latter be given in the same ratio, "in strong work" he frequently becomes "slow," and loses his action from over-exercise.

In selecting yearlings or young horses that have never been trained, the purchaser should bear in mind that the animals are on the eve of being trained, as well as at an age when rapidly growing, and requiring every care; and therefore he should have regard, not alone to the breeding and shape, but also to the condition, leaving, as stated, "something to work upon"—badly or half-fed animals being more or less in the background, in some respect or other, during their training career, and requiring double the time of others.

The period for taking up or commencing to train yearlings varies according to circumstances and opinions of owners, some commencing to break or handle them in July, others not until about October, according to their advancement in condition, time of their being foaled, as well as their early engagements. On this point it matters little, provided caution is observed in other respects. If the colt happens to have been accustomed to his paddock, with plenty of fresh air and exercise, there is nothing more likely, upon sudden change into a warm stable, than his being seized with a distemper, sore throat, swelling of the glands, frequently ending in death. Great care should, therefore, be taken to have well-ventilated stables, especially when the young animal has been suddenly changed: during the fine season especially plenty of air is indispensable, the want of it most detrimental. In the event of commencing during the summer season, while the ground is hard, great care should be taken to avoid much work, confining the exercise to walking, trotting, or gentle cantering, more with a view to making the mouth than with any other object; for if a deviation from this rule be made, the consequence too frequently is sore or "buck" shins, the bone, tendons, or muscles in animals so young not being matured: even old horses, through excessive work on hard ground, are not unfrequently affected in the same way, which renders them literally unable to move.

The enlargement in some instances is more manifest than in others; in many, soreness without enlargement, the only symptom being the shortness of the fore-action; the sufferer, in attempting to canter, going shorter than usual, without the natural freedom. The cause or even existence of it is not at all times known, even to the trainer, from the absence of the "bend," or raising of the hard protuberance in front, extending from under the knee down the fore-part of the leg, about two or three inches. The simple way to ascertain if soreness without such an appearance exists is by running the hand suddenly, with a pressure, down the front of the legs : if it does exist, the animal immediately yields with pain; in many cases, if badly affected, will almost fall on the knees. I have known horses while in training and racing to be skin-sore, and their owners and trainers did not dream of it, and have known yearlings to be rendered totally useless, and never recover their action, through this complaint; and little wonder it was so, for the parties who had charge of them (some of them their owners) could have expected nothing else, taking into consideration that they were in the month of July, on the hard ground, carrying for hours (and cantering about) great big men, their legs being little more than grizzle. Rest, a little physic, and wet applications, with bandages saturated with burned salt and boiled vinegar mixed, applied in a warm state, and left on when cold, will be found useful, giving but gentle exercise during this period. Some resort to blistering, the effects of which are more injurious than beneficial, tending to ossify instead of reducing the symptom.

Great care should be taken as to the mouthing and breaking of young horses, an experienced hand being necessary for some time; for it too frequently happens that horses' mouths are spoiled at the commencement, and, once " *made*," it by no means follows that they are *properly* made —quite the contrary. They very often become so hard and uneven on one side or the other, through having had a " bad hand" in the breaking, that they can seldom, if ever, be made as they should be—proving most injurious in their racing career, particularly on round or oval courses, such as Chester and Manchester. A steady man or boy is the proper person to select—not too heavy—who will not ruffle or "fight" with the young horse during the course of his breaking, which is not an uncommon occurrence; the consequence being loss of temper and fretting, which prevents the animal feeding during the period he most requires nourishment and strength. In short, too much kind treatment cannot be shown at this particular stage; such as speaking to, and tapping with the hand; the reverse of which leaves an impression, tending in some cases to cause vice, in others want of confidence: the latter of which is as essential as the former is ruinous. Horses, like their masters, if taught bad habits when young, seldom forget them; for, to train anything, there is nothing like bending the sapling when young and tender.

Without temper, and confidence, which usually accompanies it, a racehorse is useless. At home, in his private trials, he may be tried so highly as to convince his trainer that his coming engagements are certainties; when he appears in public, in a crowd, with the silk racing-jacket, and light, tight-fitting, racing-saddle, to which he has not been, perhaps, before accustomed, or, on the other hand, knows too well from experience, meeting a crowd of strange horses—not only strange, but equally unaccustomed to the

usual bustle and noise of a racecourse, at the very moment
when a practical demonstration of all his qualifications is
demanded, and all his powers are called in question — he is
not only rendered unable to finish and display them, but,
in point of fact, never *begins*. Hence it frequently happens
that experienced trainers win with inferior animals, beating others possessing qualities far superior to their van-
quishers. It is really astonishing the extent to which faint-
hearted horses prove not only unserviceable, but injurious
to their owners.

A curious instance of the sagacity of horses was
manifested some years ago at a country meeting. A hack-
race having been " got up," some strange animals contended. Amongst the number, one belonged to a certain
ex-member of the P. R. In the outer ring, where he had
been backing his nag, one man laid him several wagers
against him in crowns. The horse was winning the race,
and suddenly stopped at the distance. Upon our hero
making inquiries, it turned out that the layer knew he had
been in a milk-cart, and cried at the top of his voice,
" *Milk!*" The owner vowed to punch his nose !

Want of heart in the racehorse is fatal, and often caused
by harsh treatment when training. Amongst the errors
practised, a fatal one being that of trying and training
horses continually on the same ground, the very approach
to which becomes obnoxious to them. Upon this subject it
is right to observe, that when training or running at so early
a period, and for such short courses, generally about half-
a-mile, the great point to be practised is quickness in
getting off, or, more properly speaking, in racing parlance,
"jumping off," as a great deal depends thereupon. Then, in
order to carry out this principle, and bring the two-year-

old to the post, it is most desirable to teach him how to leave it, with the silk jacket, colours, and racing-saddle he is about to carry; accustoming him to crowds, noise, and bustle. In very many instances, two-year-old races are won by animals thus trained and taught; beating others in reality far their superiors. I knew a very strange instance to happen to two parties who were joined in some horses. One had, however, a two-years-old exclusively his own, which was engaged against another of the other party; a very superior animal, and backed for a heap of money, whereas the other was merely backed by his owner at extreme odds. The latter offered not to run if he received his own stake, which was declined. He started, and won in a common canter, simply because he was a good "beginner."

The object of parties seeking to win early engagements with two-years-old, should be to render them as precocious and advanced as possible; therefore, no sort of stratagem should be left unheeded to bring them to perfection. They should be ridden into market-places, fairs, and crowds of every description, as noisy as possible; a little music, in the shape of kettledrums and fiddles, will prove most serviceable for a future day. Only fancy a "green" two-years-old (many of which are frequently seen), never having seen or heard crowds, or the noise and bustle of Epsom Hill on a Derby day, or any other day; at what conclusion can that sagacious animal arrive, except that he is in danger and about to lose his life, being thereby deprived of the confidence necessary? Who can expect it at so early a period of life, to be properly developed to the extent required in the young animal about to make his *début* amongst such scenes and commotions?

The earlier you require the two-year-old the sooner you must begin to teach him that which he is about to perform; for although he may naturally possess the gift, it is the duty of the trainer to teach him how to display it, for many good ones have been lost to the owner, parted with for a mere song, while possessing qualities undeveloped, which, if placed in proper hands, might have been "stars" on the turf. Just as many eminent men have been in their youth or schoolboy days put down as blockheads. The reader must understand, that when I refer to two-year-olds, I mean from the time of being backed or put into training—say October, as yearlings—up to their appearance in public for the first time.

There are many scientific trainers who can "wind horses up" to perfection, for certain races; and whose skill is derived from extensive practical knowledge and experience. Yet there are some of these very *artistes* who almost gallop their animals to death, regardless of their nature, constitution, or breeding; for there are certain strains of blood which will not run light, or which require half the work of others. I have known instances where horses have shown good form, which induced the owners to try their hands at "higher game;" believing that, when some of the "stars" of the profession had run the rule over them, and "given them the polish," they would do wonders: the result was, that they could not beat "anything;" yet, when brought back to their old quarters, and given a few weeks' gentle exercise by the side of a hedge, some rest, a little of "Doctor Green's" remedy, in the shape of some cool green-meat, their "poor feet" attended to, and cooled also, have recovered the effects of the "skin-'em-alive" principle, and absolutely come back to their former real form; having had less work and more corn. The jaded and dried-up condition of some horses is truly wonderful—literally galloped almost to death.

A certain Irish trainer (now deceased, who in his younger days trained 'Harkaway,' 'Rust,' 'Barkston,' and many renowned horses), most experienced, but eccentric in his character and manner, many years ago trained for me and others. He was remarkably averse to over-drawing horses, and especially fond of the 'Birdcatcher' blood, which, he very properly remarked, did not generally require as much work as others. Nothing used to annoy him more than my wishing to have horses " rattled along." As for " trials," it was next to an impossibility to get him to consent. If I called to the boys, when at exercise, to " go along," he would reply, " *Oh, blur an' ounds, you'll burst them !*" at the same time raising his hand and roaring aloud, " *Hould hard !*" And turning to me would add, " *Tare an' agers, this'll never do !*"

Having afterwards become private trainer to a certain deceased and lamented nobleman, who, in common with all who knew him, entertained a very high opinion of him, and being a first-rate horseman across country, a certain celebrated steeple-chase was about to come off, in which he was to ride a renowned horse, a great favourite. The night previous to the chase his lordship, who happened to be stopping at the hotel in the neighbourhood of the course with some friends, was startled by an announcement that a messenger in great haste wished to see him on urgent business; and having asked him the nature of his errand, in a most excited manner the latter exclaimed, " *Och*, my lord, there's *murther* down the town ! Larry" (the trainer and intended rider) "has had a *dhrop* too much,

and has been *wallopin'* some *Peelers* with a flail." "Has
he been taken?" "Faith, not he, my lord; he got his
back *agin* a wall in the barn, and stretched them as they
*cum* on." Larry, nothing daunted, and regardless of the
risk of being arrested by the *Peelers;* who were on the
look-out for him, being.determined to witness his horse's
running next day, repaired to the course in female disguise, and from the top of a corn-stack witnessed his success. He used to take the pledge for a *year and a day,*
and on the expiration thereof no Father Mathew could
prevent him having a " spree;" during one of which, in
the time of the famine in Ireland, he commenced a letter
to me,—" Awful Sir;" and addressed it at bottom, " Awful
Times, Esquire."

One of his principal objections was to small horses.
Upon one occasion a horse of mine, ' Chief Baron,'
(brother to ' Micky Free,') had just left the starting-post,
amongst others, against a very large mare called ' The
Baroness' (sister to ' The Baron'), belonging to Mr. Watts.
As we were riding across the course towards the winning-
post he remarked, looking back towards them, " She'll
*murther* the poor little fellow with that *awful* stride." He
was, however, agreeably surprised to find old ' Denny
Wynne' win by a head, to the dismay of Mr. Watts;
who remarked that *his colours* won: at which Larry remarked to me, " Tell him they're yours, too."

Many horses are overloaded with clothing, and kept
in perfect hot-houses; the results being most injurious
in many respects. Amongst others, from causing liability
to colds and distempers. For horses engaged, or about
to run very early, and especially to those inclined to
long, or heavy " coats," there is nothing better than

the system of clipping, or removing the winter coat. It improves their condition in many ways, and they thrive much quicker.

Horses are frequently too much hurried in their training, through the anxiety of owners to run them for certain engagements; the consequence being most injurious. Many owners fancy they can be got fit to run in half the time really required. A young beginner on the turf having sent a horse to be trained, with instructions that he should be *ready in a month* to fulfil a certain engagement, although at the time a mountain of flesh, upon the eve thereof visited the trainer's establishment, and having been shown round the stables, requested to see his own horse, as he was in a hurry to catch the train — he happened to be in his horse's box at the time. " That is your horse, sir !" remarked the trainer. " What ! do you mean to tell me I am a fool ? That my horse, indeed ! Where is my horse ? That is not the one I sent you !" The trainer reiterated the assurance that it was; the owner, in a fit of laudable indignation, exclaimed, " Why, sir, that's not the half of him !" and adding, " that he had better secure the remaining half while he could," immediately removed him.

It is quite true, that the preparation for such a race as the Derby tests to the utmost the skill and intellect of the trainer; the object being, in most cases, to bring off that event upon the " winding-up " principle : or, in many instances, risking the ruin of the animal; as if his services never would be required again. The consequence is, that more valuable horses are broken down from this cause, when training for this race, than many of the others put together. " Such a horse will never stand a Derby pre-

paration!" Why? Because he is frequently galloped
to death, and the owner, on the eve of the event, politely
informed that his horse has broken down. No doubt
the Derby course is about the most severe and trying
in England, for legs, condition, &c. But that fact
should be especially borne in mind; and while studying
the muscular condition of the frame, the preservation of
the legs should not be forgotten, which frequently happens.
It is quite true, that to bring the former to perfection
good legs are indispensable. "The trainer naturally
says, You must give me good-legged ones." How many
Derby winners are never afterwards heard of, except at
stud!

The manner in which some animals vary in form,
during even one season, is extraordinary, yet can be accounted for in various ways. Like every other animal,
they are not at all times well, and in the same vigour of
health, from various causes. Sometimes, through excessive work; at others, through simple, perhaps slight distemper, under which they are frequently labouring, even
when running races : yet it does not absolutely break out,
or appear visible, perhaps, until the day after the race.
Moreover, many "mares" towards the end of the season,
when casting their coats, become weak, and do not show
their real form for a certain period; just as fowl become
weak when moulting.*

* It is believed by most people that what is termed "turning
horses up," or taking them out of work, injures and lessens
their action. That is a point admitting of doubt. There have
been instances where horses and mares have won, and over long
courses, after being put to stud. A stallion, called 'De Vere,'
many years ago, and a superior horse, won cups and Queen's
plates, four miles, after being at stud; and, if I mistake not,

Some legs last much longer than others, and in many instances *doubtful-looking* ones prove the best.

When purchasing yearlings and entering them for engagements, a great deal should depend upon their shape, size, breeding, and the distance of the courses ; the great mistake, in my opinion, being that they are frequently too heavily engaged : for any experience I may have had in such matters, induces me to recommend owners to wait until they know whether they are worth engaging, at least, before entering too deeply. Moreover, not only the accidents, but diseases and chances to which young horses are subject, are numerous, and only known to those who have paid for their experience : for instance, take the late Lord George Bentinck's entrances for the Hippodrome stake — eleven in number — seven of the lot died before the day. Another instance of the " glorious uncertainty," and more especially as regards hazard of " matchmaking," was borne out in the case of three yearlings that had been matched against each other to run at two years old. Having matched a yearling of mine, ' The Maid of Moorfield,' (by ' Magpie,' out of sister to ' Irish Birdcatcher,') against one of the late Marquis of Waterford's, for five hundred sovereigns a-side, he naming a filly by ' Coranna,' out of ' Repartee,' also matched against the late Lord Caledon's brother to ' Shylock,' for a similar amount, the result was as follows :— The last-named was shot in his paddock shortly afterwards, by some recently discharged

for more than one season. And I have known mares to win Queen's plates, four miles, within three months after foaling, the produce being reared by hand. It would be well for many if they were turned up occasionally ; for, in my opinion, it would bring their lost action back, at least in some instances.

servant; Lord Waterford's filly broke her back a fortnight before the match; and mine died within a few days thereof, of distemper.*

Although I write against the prudence of matchmaking, it is rather curious that in such I proved more fortunate than in other engagements, having made a good number, and never lost but upon one occasion. Still, to run for public money is more judicious, profitable, and less precarious, except when horses are fit and well, when it is probably as desirable as any course.

There is no age at which horses can give so much weight to others as at two years old; the earlier they are tried the greater the disparity will appear: the causes can best be traced to the following reasons :— The difference as to size, shape, power, speed, breeding, early condition; in short, *artificial* as well as *natural* advancement in the respects required. Take, for instance, such contrasts as are frequently witnessed, where some of those speedy flying fillies, well trained, and taught to "jump off," not only by nature, but through prac-

---

\* It would indeed be an act of ingratitude to omit mentioning the noble and generous manner in which his lordship acted upon the occasion, as he did upon other similar ones with me. His filly had just run in a race, and although a very small, and, indeed, inferior animal, still she was sound and well within a fortnight of the match. I asked, "What will you give, my Lord, to be off *now?*" To which he replied: "Lord Caledon's is dead; is yours yet, as I know she is, near it?" "Very nearly, my Lord. Then what will you take?" "I'll leave it to yourself. Anything you like." I handed him fifty pounds: it was half forfeit. "Now," said he, "I shall run this mare again." He did so, and she broke her back. Mine died within three days of the match, — the best filly I ever tried: better than 'Early Bird;' the same age, and as nice a mare as ever was looked at.

tice and attention in the way of training and trying, so
frequently bestowed by some trainers for early engagements, the distance being half a mile or under; let them
be opposed by others—fine, slashing, overgrown horses
—what chance can the latter have, as a general rule?
The chief cause, taking two-year-olds in the general run,
is simply that some are so much better reared, fed, and,
in fact, forward in every respect.

Horses, during their racing career, arrive at their best
form, and display their powers in their greatest perfection,
according to their breeding, shape, the period at which
they have been in strict training, and the amount of
work they have done. Some are of an improving breed,
and "train on;" others are "fliers," at two and three
years old, and as they grow old, get worse—no better,
certainly, after three years old. But to my mind many
of the regular-class racehorses, trained and treated in
a fair way, according to the general run of the first
class in the present day, are nearly, if not quite as
good, at three years old in October, as ever they are, say
for two miles: that is to say, if they have been racing
much at two and three years old, and provided they are
kept at racing weight in October. This remark does not
apply to animals of certain blood, that may have been
judiciously kept from severe work at two years old, and
have been kept over; because any animal if thus treated
must improve, and be better at four or five years old. Of
course there are exceptions, especially where there is fine
size, substance, power, &c.: but my remarks are with
reference to horses as they are generally worked in the
present day, the effects of which tell so that I believe
"the clock has been wound up to te last link of the

chain," as a general rule, in October, at three years. The "stamina" may increase, but the "speed" begins to slacken; indeed, many Derby and St. Leger winners are never afterwards heard of except at stud. Amongst others, ' Faugh-a-Ballagh' and 'The Baron' furnished proof of the fact.

Many hasty conclusions are formed as to the deficiency in staying powers of certain horses. They are sometimes considered "half-milers," because they have been *tried* and *trained* for *that distance*, and having shown speed, are frequently set down as "non-stayers," engaged accordingly, and because they win at such a distance the "name sticks to them;" like the man, who had the name of "getting up very early in the morning, yet used to remain in bed all day." The consequence is, that because they have shown speed, they are considered animals without staying powers. I have known instances, and have myself purchased horses from "flash stables" where they were thus condemned, and had run solely for half-mile races; and yet their *forte* was a distance, and four miles would have suited them better than half a mile: they won over two miles and upwards, beating good horses, very much to the astonishment of their previous owners and trainers.

If horses naturally possess speed, it is wonderful what time and training will accomplish; as to staying, if they have it not, no time or care can place that gift where it has been denied by nature. The large, perhaps leggy, overgrown two-year-old, even if he can run but a few hundred yards, will with time make wonderful progress as to staying. At the same time it invariably follows, that where animals are possessed of speed to an extraordinary degree, they are invariably deficient in staying powers, their shapes as well as action being different to those of stayers; indeed, to a certain extent, the two qualities, if shown in any extraordinary degree, are incompatible, although the latter may be gifted with moderate or "good" speed. My remarks refer merely to those very short, compact, half-mile fliers.

If the ashes of Alexander, or the cruel Caligula, were to rise from the grave and witness the wonderful exploits of some horse-tamers of the nineteenth century, how they would stare! There can hardly be a question that the iron-bondage system would be resorted to, and the practitioners asked how their joints felt after a month's trial. Probably, the sentence might be a tar-barrel rolled from the top of the hill towards the starting-post at Epsom.

Even the Zebra could be tamed by domestication and proper treatment, in time; but he has not been considered worth the trouble. One generation might not prove sufficient. Like the lion-tamers and performers, it might also astonish the reader to learn that the *real* tamer could be found in a very humble situation as ostler, not one hundred miles from Brompton, London; his implement being a hot iron bar.

The horse appreciates kind treatment, when shown to him, far more than his ungrateful master does. Take, for instance, the trained circus horse: does he not perform according to the dictates of his trainer? Ought not even this simple fact be sufficient to convince any party that kindness on the one hand, and, when necessary, chastisement on the other, should be observed? Then, as before stated, the time to commence is when they are young. The ill-

M

treatment to which they are sometimes subjected frequently leads to vice and bad temper. Surely such beautiful creatures were not formed to be " walloped," as his lordship's Larry did the " Peelers?"

That noble animal the racehorse, generally and so justly prized by man, although so cruelly treated by some, is docile by nature, and more capable of appreciating kindness than many of those who turn him to mercenary and useful purposes. There is more gratitude in one hair of a horse's tail than in half mankind. He is noble, not alone in appearance, but by nature and instinct. His perfections and superior qualities are seldom so clearly discovered and confessed as when, in his majestic form, he stands the scrutiny of the admiring crowd upon the occasion of " weighing in;" when, as far as a dumb animal can give expression to his feelings, whether of gratitude or contempt, he gives his backward kick, as much as to say, "That for your opinions! where is the corn-bin?" His symmetry, his muscular development, his beautiful head, eyes of flashing fire, heated veins, and extended nostrils, then call forth most laudable ebullitions of admiration from spectators; while the owner calmly calculates the addition to his coffers by the victory so gamely struggled for and gained. But unfortunately, in too many instances, that noble steed, that once, when in bloom, had so well served his master, serves also as a sad example of the latter's ingratitude; and having reaped many laurels, like the faded flower withered and forsaken, is cast from the breast of the spoiler, and replaced by a blooming rosebud—forgotten until, perhaps, at some future day, again recognised by his former master when safely lodged at a railway station in a Hansom, on his journey to a race-

meeting.  So unlike the testimony of attachment placed by
his owner, Sir Gilbert Heathcote, in memory of ' Amato,'
or Orlando to his dying horse, his faithful steed that long
had served him well,—

> "' My much-loved steed, my generous friend,
>     Companion of my better years!' he said,
>   ' And have I lived to see so sad an end
>   Of all thy toils, and thy brave spirit fled ?
> O pardon me if e'er I did offend,
>   With hasty wrong, that mild and faithful head.'"

THE

# SIRES OF THE PAST AND PRESENT DAY.

———

" The tainted branches of the tree,
   If lopped with care, a strength may give,
   By which the rest shall bloom and live,
  All greenly fresh and wildly free."

As the subject of Sires appears to occupy so much of the
attention of the sporting public, and as even deputations
wait upon the Government with the view of devising means
of improving the breed of horses in general, I venture to
make a few remarks thereupon.

We find that there are now some hundreds of stallions
at the service of the public in Great Britain and Ireland.
To deny that the greater portion are descended from
first-class blood would be absurd; yet there are very many
other necessary qualifications which demand the attention
of breeders, according to the object they may have in
view—whether it be to become possessed of a useful
animal for general purposes, or, on the other hand, to
run that well-known risk which admittedly accompanies
every endeavour to breed a racehorse.

Allowing that there are many whose blood is equally
fashionable and faultless, still there are some which may
possess extra recommendations as to size, strength, shape,

colour, and performances, either during their racing career or as stud-horses.   And when discussing the merits of *tried* sires, it by no means becomes so necessary to look to their own merits as racehorses, as to those of their sons and daughters; for many inferior racehorses have proved good sires, and *vice versâ*.   Still the chances must be naturally in favour of those as near perfection as possible, and with the least drawbacks.   A very great difference of opinion exists, in the present day, with regard to the staying powers, speed, and other qualifications of certain strains of blood; the partisans of each, no doubt, advancing strong arguments in favour of their respective views, although occasionally in a rather prejudiced manner: without taking into consideration the various circumstances which may have tended towards increasing or diminishing the *prestige* of each.   Such, for instance, as the number and quality of the *brood mares*.   As regards the former, it by no means follows (even if the animal should be standing at fifty guineas, and his subscription full) that he has had a better chance than another with perhaps half the number, for the following amongst other reasons : viz. every owner of a good brood mare is not at all times so flush in funds as to place him in a position to pay that sum of fifty guineas, together with the incidental expenses of travelling, &c.   Whereas there are many wealthy noblemen and gentlemen, whose purses or properties are more extensive than their know-ledge of horseflesh, who at times keep a very inferior class of brood mares, and who, from some whim or fancy (perhaps the mare having on some previous occasion jumped over a donkey's cart, a five-barred gate, or carried her ladyship brilliantly with hounds), do not hesitate to pay the large figure; the consequence being, that the sire is believed to have had chances which he really never had; while another, equally well bred, and otherwise desirable, has been patronised with a lesser number: the property, perhaps, of experienced judges, as well as parties who pay every attention, and leave no stone unturned to bring to perfection the produce. The consequence is, that the casual reviewer of the statistics, generally recording the number of foals, winners, &c., forms a hasty conclusion as to their respective merits and chances.

We frequently see instances, where what are termed *unfashionable* sires have produced first-class racehorses; the fact being, that they have been *unpatronised*, although, in many instances, their superior blood, as well as first-class performances, entitled them to it: amongst others may be mentioned 'Ivan,' a horse of undoubted blood, as well as a first-class racehorse, as his performances with 'Vindex' and others will testify. Besides, we have 'Van Galen,' 'Syphon,' another 'Van Tromp;' a striking proof of the value of the 'Lancrcost' strain, especially for stoutness, evidenced in their respective sons, 'Union Jack' and 'Tim Whiffler.'

There can be no greater proof of the lottery of breeding than the success or failure of sires; for no matter of what blood, or how they may have distinguished themselves as *racehorses*, still, in many instances, the most signal failures and disappointments are experienced, where the performances. as well as high breeding, would lead one to expect most successful results: for instance, 'Cotherstone,' 'Pyrrhus the First,' 'Charles the Twelfth,' 'Surplice,' 'Launcelot,' &c.: yet they sometimes produce *one* or *two first-class racehorses*: for instance, 'Cotherstone,' sire of 'Stilton,' and 'Pyrrhus,' of that extraordinary mare

'Virago' (whose trainer believed her to be about ten pounds better than 'Crucifix'). In fact, the failure of such horses as those mentioned, with all the chances they have had at stud, is one of those things which, as Lord Dundreary says, "No fellah can understand," still it may to a great extent be accounted for, in various ways : the most frequent and fatal mistake being too much "stall-feeding," and insufficient exercise, the groom not unfrequently sitting over the fire, smoking his pipe, when he ought to be exercising his horse. However, it is perfectly absurd to think that any man can properly attend to the number of animals that some owners fancy they can. Some grooms have about three times more to do to than they should have. To attend properly to one sire is enough for any man, during the season and while preparing for it. It is almost impossible to give a stallion too much exercise by hand; yet it must be very hard work walking for hours and leading a restless sire, which is much more judicious exercise than that in the ring.

If possible, the best way a stallion could get his work, or exercise, is simply by being ridden; and in summer, when done his season, let him have his paddock to exercise in (although I am aware many will differ in this view, on various grounds) : but, with all deference to those who adopt the "stall-feeding" principle, even in the dog-days, my humble opinion is that the large, loose, well-ventilated box and well-enclosed paddock, would prove more beneficial, and would tend much to save the lives of many that die from inflammation, brought on by confinement and want of proper exercise. How frequently do we hear of sires becoming roarers *after* being put to stud, where even the very suspicion of such an infirmity never existed, up to the end of their racing career: and why? Simply because one-half do not get sufficient exercise, whilst some get none at all, at any period of the year, but are turned up and overfed. Taking into consideration the fact that such animals have but recently ceased not only natural, but artificial and overstrained work, it is hardly to be wondered at that the sudden cessation of such exercise, as well as the different and fattening nature of the food which they receive, should tell its tale. But from whence do such reports derive their origin? From servants sent with mares, who inform their masters that "they heard the horse roar;" which, in point of fact, was nothing more than a grunt like swine, produced by sudden exertion and from being so full of fat inwardly, which in many cases would gradually disappear if the horse were again brought back into his former racing condition. Another most injurious consequence of want of exercise and over-condition is the *uncertainty of produce;* for there is no doubt that the most certain foal-getters are the common travelling stallions, who are engaged almost the entire day in walking from place to place. Many instances have been known, and come within my own knowledge, where mares, barren for *years* to other "made-up" sires, have had produce *the very first time* they were served by this class. A gentleman some years ago purchased a young stallion from me (brother to 'Shylock') at four years old; the horse had never been tried while in my possession, or to my knowledge previously. He had a large number of mares subsequently, but none proved with foal. Yet, as well as I remember, the horse being afterwards put to work during certain periods of the year, turned out fruitful.

A similar case happened with a young prize bull which

I purchased at a cattle-show for a large price. Being housed and kept in high condition, out of nearly fifty cows, there were but two or three calves. When turned out he was quite fruitful.

Many persons fancy that the cause of roaring is a thickness or swelling of the glands or muscles of the throat, whereas it is exactly the contrary. It is the wasting or withering away on the one side thereof, the other remaining in its natural state. If any one consults a competent veterinary surgeon on the subject of a horse's wind, even if the animal be in light condition, he will have the horse about to be examined, if possible, "galloped" for a considerable time, before he can arrive at a *positive* conclusion as to his soundness. It is hardly likely, therefore, that persons can easily come to a decision as to the soundness of a stall-fed stallion. Some of the soundest, best-winded, and longest runners, make a sort of snorting noise through their nostrils, even when going half speed. The principal reason why some owners keep their horses in high condition is, because they very naturally believe they look better to the eye of most people; so they do—and the occasional "flatcatchers," in the shape of fancy head-collars, &c., have their effect in the eyes of many. Few men know how an inch off a horse's tail can add to his substance and lessen his height, &c. A certain noble Lord, B*l*d*re, once told my grandfather "he did not like a certain horse's tail." "Oh!" replied the latter, "it is not on his tail you ride!"

If the breeder has plenty of capital, and can afford to speculate, he could not do better than pick up some extraordinarily well-bred young one, of good shape and promise —suppose a three-years-old, that has been unsuccessful as a racehorse, if let out at a small figure, or even gratis, to *tried* mares, for one season or two, might in time establish for himself a name worthy of his ancestors: for instance, such a horse as 'Petruchio' by 'Orlando,' out of 'Virago.' This animal, if he has grown into what he promised when a yearling, with fine propelling power, and good-looking enough for anything, would be the sort to select, "for blood will tell." It has been said by many experienced judges, that speed is the great point to look for in the *brood mare*, and it is to be presumed the same rule ought to apply to the sire. Whether this opinion is correct, and I for one concur therein (inasmuch as however time, training, and good management can make horses "train on" and stay, no power or skill can put *speed* into a horse unless he *naturally* possesses the gift), yet there are, and have been, very many instances where mares having proved worthless as racehorses, still have turned out tip-top brood mares; and there is no reason why it should not prove the same with stallions, provided always they have the shape, soundness, size, and above all other recommendations, "*that they belong to a running family,*" although perhaps, from some cause or accident, they may have proved of no value on the turf.

There are many extraordinary instances in the lottery of breeding, and perhaps none more so than where we find such horses as 'Cotherstone,' 'Coltsterdale,' and others, although racehorses, comparative failures as sires; and yet their own sisters, 'Mowerina' and 'Ellerdale,' not only good but first-class brood mares: for example, they not only produced each a really good animal, but absolutely proved successful with many strains of blood, producing such animals as 'West Australian,' 'Ellington,' 'Gildermine,' 'Wardermarske,' 'Summerside,' 'Ellermire,' &c. Must not this fact go a great length towards proving that the good green pasture and exercise served the dams, and that the dried-up system observed with the sires had the contrary effect?

Taking into consideration that such a horse as 'Petruchio' (bred as he is, with probably the most high-sounding pedigree of any horse living) did not turn out a racehorse, do we not find a fact more to be wondered at than if he should hereafter prove the sire of first-class horses, if he should get the chance; particularly bearing in mind that he has gone to stud a fresh young horse, with constitution unimpaired by training, and probably otherwise sound? As far as this animal is concerned, I know nothing as to whether he has had any chance as yet, nor anything further regarding him; I merely cite him as an example. If we take the winners of the three great events, viz. Derby, Oaks, and St. Leger, for the last thirty years, and contrast their success at stud, either as sires or brood mares, the results will prove anything but encouraging to breeders—with very few exceptions indeed. And it is also worthy of remark, that the winners of the St. Leger have proved far more successful as sires than those of the Derby; while the Oaks winners have likewise, with very few exceptions, been perfect failures: although, as a natural consequence of their success for these events, they had far better chances of distinguishing themselves than were offered to others.

| Derby. | Oaks. | St. Leger. |
|---|---|---|
| 1834. Plenipotentiary. | Pussy. | Touchstone. |
| 1835. Mundig. | Queen of Trumps. | Queen of Trumps. |
| 1836. Bay Middleton. | Cyprian. | Elis. |
| 1837. Phosphorus. | Miss Letty. | Mango. |
| 1838. Amato. | Industry. | Don John. |
| 1839. Bloomsbury. | Deception. | CharlestheTwelfth |
| 1840. Little Wonder. | Crucifix. | Launcelot. |
| 1841. Coronation. | Ghuznee. | Satirist. |
| 1842. Attila. | Our Nell. | Blue Bonnet. |
| 1843. Cotherstone. | Poison. | Nutwith. |
| 1844. Orlando. | Princess. | Faugh-a-Ballagh. |
| 1845. Merry Monarch. | Refraction. | The Baron. |
| 1846. Pyrrhus the First. | Mendicant. | Sir Tatton Sykes. |
| 1847. The Cossack. | Miami. | Van Tromp. |
| 1848. Surplice. | Cymba. | Surplice. |
| 1849. The Flying Dutchman. | Lady Evelyn. | Flying Dutchman. |
| 1850. Voltigeur. | Rhedycina. | Voltigeur. |
| 1851. Teddington. | Iris. | Newminster. |
| 1852. Daniel O'Rourke. | Songstress. | Stockwell. |
| 1853. West Australian. | Catherine Hayes. | West Australian. |
| 1854. Andover. | Mincemeat. | Kt. of St. George. |
| 1855. Wild Dayrell. | Marchioness. | Saucebox. |
| 1856. Ellington. | Mincepie. | Warlock. |
| 1857. Blink Bonny. | Blink Bonny. | Impéricuse. |
| 1858. Beadsman. | Governess. | Sunbeam. |
| 1859. Musjid. | Summerside. | Gamester. |
| 1860. Thormanby. | Butterfly. | St. Albans. |

With the exception of the success of the 'Whalebone' strains, very few of the above have proved valuable or successful as sires, and most of the mares have almost invariably been useless.

In choosing the blood to commence with, on the sides of sire and dam, it would be absurd to be prejudiced to too great an extent, either one way or the other; although the breeder for sale must, in a very great measure, *go with the times and fashions of the day;* and in attempting to pick out a favourite strain of blood, with so many to choose from, one, to a great extent, becomes (like a lady selecting a dress or a bonnet) bewildered, and unable to fix upon any particular choice.   However, I shall, without favour or prejudice, express my humble opinion on the blood and crosses generally, and lay before the reader my ideas.   But before doing so, I beg to express my opinion with regard to the course which a breeder should take, as to keeping his own sires or hiring those of others.   To my mind it is an error for an owner of a lot of mares to tie himself to any particular stallion or stallions; if he keep his own, he is frequently led away, by prejudice perhaps, to sacrifice too many of his mares in the hope of establishing the sire; and year after year he may repeat this course, which, although it may turn out successful, still is increasing the risk of breeding to too great an extent: especially as he may be damaging the *prestige* of his brood mares while trying to accomplish his other aim—thus undertaking, in fact, a mere risk as to profit or loss, between enhancing the value of the sires or mares, both of which, quite possibly, might prove successful. Although we find instances of own brothers and sisters turning out well, such as 'Whalebone,' 'Whisker,' 'Wire' and 'Web,' 'Stockwell' and 'Rataplan,' 'Touchstone' and 'Launcelot,' 'Mountain Deer,' 'Sylphine,' 'Champagne' and 'Claret,' as well as 'Irish Birdcatcher' and 'Faugh-a-Ballagh,' besides the late Mr. Watts's lot, the produce of his little mare 'Clari,' viz. 'Chat,' 'Chatterer,' 'Chatterbox,' 'Chitchat,' and 'Third of May,' own brothers, still there are very many counterbalancing instances, where the dam has produced very indifferent animals—own brothers and

sisters: take, for instance, 'Vanderdecken,' brother to
'The Flying Dutchman;' and yet, when she was changed
to 'Lancrcost,' she produced that extraordinarily good
horse 'Van Tromp:' 'Crucifix,' dam of 'Surplice,' produced 'Pontifex,' an own brother, who turned out badly,
and when put to 'Lancrcost' she no doubt failed to
improve the quality of her produce in 'Crozier.' There
arc numerous other instances of the kind.

Before referring to the merits of modern sires in
particular, let us glance over those of some of their
ancestors, without unnecessarily tracing back too far.
Let us commence with the renowned 'Whalebone' blood,
in favour of which so great a prejudice naturally exists
in the minds of breeders and purchasers; a prejudice
which, no doubt, has its origin in the fact, that it furnishes more winners, and consequently, with more speedy
returns, brings more " grist to the mill " than any other,
whether at sale as yearlings or tried subsequently: but,
to a very great extent, such successes arc attributable to
the vast number of the scions of this family who contend
for races, and who arc so extensively patronised in other
respects, for upon reference to the catalogue of sires at
present at the public service, we find that on one side or
other, within two generations, three-fourths can claim
near relationship to this blood, some being, on either the
side of sire or dam, a 'Birdcatcher' or a 'Touchstone,'
while at the same time other valuable strains are almost
discarded *in toto ;* a state of things leading to the supposition that we may shortly expect to find breeders
adopting a system, in respect to some blood, approaching
the old Egyptian custom in the days of Cleopatra, while
some appear likely to share the fate of the Kilkenny cats.

Before, however, entering into the respective merits of the two extraordinary animals referred to, 'Touchstone' and 'Irish Birdcatcher,' whose relative merit as race-horses or at stud it would be difficult for any unpre-judiced person to decide on, a retrospective glance at 'Sir Hercules,' the sire of 'Irish Birdcatcher,' and probably 'Whalebone's' most valuable son, may not be out of place.

'Sir Hercules,' by 'Whalebone;' dam 'Peri,' by 'Wanderer,' 'Alexander,' 'Rival,' by 'Sir Peter,' &c. The extraordinary merits of this horse, as a sire, hardly require comment. It may not, perhaps, be generally known, that he was forgotten for many years, and left to serve half-bred mares belonging to farmers, at Summer-hill, in Ireland, at a figure in amount about equal to the groom's fee chargeable for other sires; and to this day there is hardly a half-bred colt sold at Irish fairs that the owner will not trace back to '*Ould Sir Hercules.*' He is certain to be by 'Young Hercules,' or his dam by the "*ould horse,*" particularly if he has the grey or silver hairs in his tail, or on his quarters or flanks, which is invariably the case where there is a drop of the 'Sir Hercules' blood. His son, the renowned 'Irish Bird-catcher,' shared a similar fate, having—although let out at a very small figure—obtained little patronage; and were it not for the liberality displayed by his owner in reducing the small figure at which he stood, he would not even have paid the expenses of his keep, although his stock had won repeatedly in Ireland; and he might have been completely lost to the country had he not been hired to serve in England. From 'Sir Hercules' are descended in a direct line the following sires, besides

many others : — 'Irish Birdcatcher,' 'Faugh-a-Ballagh,'
'Lifeboat,' 'Gunboat,' 'Gemma di Vergy,' 'Stockwell,'
'Rataplan,' 'Saunterer,' 'The Marquis,' 'Leamington,'
'Ethelbert,' 'Big Ben,' 'Wonnersley,' &c. His colour
black, with grey hairs in his flanks and tail, appears to
be handed down to his son 'Saunterer,' a horse in other
respects very much resembling his grandsire, and one
most likely to prove a very great loss to his country, for,
although he was not a very large horse, he was of that
caste and metal best described as "steel : " all bone,
sinew, and muscle — no lumber.

'Touchstone,' as everybody knows, during his racing
career, proved himself an undoubted racehorse; defeating,
amongst others for the St. Leger, the Derby winner,
'Plenipotentiary' (at the time surnamed 'The Lion of
Doncaster'). His numerous successes at the stud require
no comment, further than a reference to the Racing
Calendars; as they would be too numerous to detail,
except by mentioning a chosen few of his sons and
daughters: such as 'Cotherstone,' 'Orlando,' and 'Surplice,' winners of the Derby; 'Mendicant,' winner of the
Oaks; 'Blue Bonnet' and ' Newminster,' winners of the
St. Leger ; as also the following good animals — ' Moun-
tain Deer,' 'Champagne,' 'Sylphine,' 'Claret,' 'Rifle-
man,' 'Typee,' ' Lord of the Isles' (winner of the 2000-
guineas stake), ' The Marionette,' 'De Clare,' 'Ithuriel,'
'Adamas,' 'Ambrose,' and ' Annandale ;' besides his
grandsons and granddaughters, 'Teddington,' 'West
Australian,' 'Beadsman,' and 'Musjid,' winners of the
Derby; 'Iris' and 'Marchioness,' winners of the Oaks;
'Impéricuse,' 'Gamester,' 'The Marquis,' and ' Lord
Clifden,' winners of the St. Leger; ' Fazzoletto,' 'Long-

bow,' 'Marsyas;' and though last, not least, 'Dundee.'
The generality of this horse's descendants are in colour,
almost without exception, dark, or rich bay or brown;
with a very few black, or a very peculiar dark mealy-
brown, with a white spot, sometimes blaze, on the forehead and nose; and more or less white about the legs:
the shoulders being low and muscular, presenting an appearance of roundness or heaviness (which many of the
'Orlandos' also display), with length, especially behind
the saddle; muscular and powerful loins, well-arched
quarters and thighs, well-turned under, with clean hocks,
denoting great propelling power; yet with a flatness and
shallowness of the ribs, especially the back ones, and
hollowness between the latter and the hips, and an un-
usual display of the white portion of the eye.

'Irish Birdcatcher,' the son of 'Sir Hercules,' and
'Guiccioli,' by 'Bob Booty,' out of 'Flight,' by 'Escape,'
like his rival sire, could take his part on the racecourse
as well as in the harem.

"When Greek meets Greek, then comes the tug of war."

The wonderful performances of this animal are unknown to many supporters of the turf in the present day;
almost forgotten by others who have witnessed them —
a reference to "The Racing Calendar" is recommended
to the reader, should he wish to arrive at a conclusion as
to the respective merits of these two horses. The extraordinary speed of the animal in question reminds me of
an incident which happened during a race at the Curragh,
many years ago, in which the best horses of the day contended. A trainer having stationed himself within half-
a-mile of the winning-post, in order to watch the progress

N

of the race, and, if necessary, give his jockey instructions —upon seeing the latter pass him, several hundreds of yards in the rear, shouted at the top of his voice, "*Purshue him! Purshue him!*"—pursuit at the time being hopeless.

But, before entering into a detail of his brilliant career as a sire, it is but right to observe that had both he and 'Touchstone' being favoured with what is so frequently *heard of* at a hazard-table, as *equal main and chance*, it would be a nice point to decide between them as to merit, in any shape or form; for it must not be forgotten, that while the "Bird" was trying to "pick up his crumbs" in his own country, with very little chance of success, the old "Gem" was being patronised by the "pick of the harems" of all countries, and reaping a rich harvest therefrom; his subscription being full almost before publication: but whatever their respective merits may have been, it is quite certain

"No future day will see their names expire."

Before referring to the descendants of 'Birdcatcher,' it is but just to observe that, with regard to England and the great "events," the Derby, Oaks, and St. Leger, 'Touchstone' had about ten years' start, as to chances. If 'Launcelot,' his own brother, won the St. Leger in 1840, the respective merits of the two animals in question are, as far as 'Birdcatcher' is concerned, by no means lessened by the success for the same race of 'Faugh-a-Ballagh,' in 1844; nor yet by that of 'The Baron' in the following year. Then, in 1852, we find his son, 'Daniel O'Rourke,' winner of the Derby; his daughter, 'Songstress,' winner of the Oaks; and his grandson, 'Stockwell,' winner of the St. Leger. In 1854, his son,

'Knight of St. George,' winner of the St. Leger. In 1856, his son, 'Warlock,' winner of the St. Leger; and his grand-daughter, 'Mincepie,' winner of the Oaks. In 1858, his granddaughter, 'Sunbeam,' winner of the St. Leger. In 1860, his great-grandson, 'St. Albans,' winner of the St. Leger. In 1861, 'Kettledrum,' winner of the Derby; and 'Caller-Ou,' of the St. Leger. In 1862, his grandson, 'The Marquis,' winner of the St. Leger; and in 1864, the past year, his great-grandson, 'Blair Athol,' winner of the Derby and St. Leger; besides the following good animals—his sons and daughters: 'Cawrouche,' 'Chanticleer,' 'Early Bird,' 'Saunterer,' 'Exact,' 'Habena,' and others too numerous to mention; not forgetting 'Justice to Ireland,' a much better animal than he was generally supposed to be, his powers (like many of Erin's other productions) not having been fully developed.

The colour of 'Irish Birdcatcher's' stock is principally chestnut, occasionally bay, with grey or silver hairs on the flank and tail, with the same white marks as the 'Touchstones:' their fault in shape for racing purposes being sometimes deficiency in length, with occasionally a slight inclination to curby hocks. They possess plenty of substance and symmetry all over, being extremely well "ribbed up," with very little space from the back ribs to the hip; very good behind the saddle; many of them, like their sire, however, being rather flighty in their temper: but it is an undoubted fact that they could all "run a bit," and were always remarkable for extraordinary speed and quality.

While upon the subject of the 'Whalebone' blood, a few remarks upon the descendants of his own brother,

'Whisker,' may be admissible. Who that remembers his grandson 'Harkaway' (by 'Economist'), his brilliant performances as a racehorse, will hesitate to admit that it is a very nice point indeed, as to whether we have since seen his superior on a racecourse, although, like most " nine-day wonders," his victories have been comparatively forgotten in the anxiety of parties to patronise some of the " mushrooms" of more recent date? 'Harkaway' had more of the 'Godolphin' blood in his veins than, perhaps, any other sire of his day. Luckily, there is at the service of the public a splendid son of a renowned sire, likely to sustain in every respect the *prestige* of the old " King of the Curragh," namely, 'King Tom.' A greater confirmation of the value of the 'Economist' line and the double cross of the 'Whalebone' blood we can hardly find, than in the case of 'The Baron' (sire of 'Stockwell' and 'Rataplan'). 'The Baron' by 'Irish Birdcatcher,' by 'Sir Hercules' by 'Whalebone,' his dam 'Echidna,' by 'Economist,' by 'Whisker' (own brother to 'Whalebone'), bears out the truth of the remark I have so frequently heard made many years ago by the " old heads," "that there was no cross like the double one of the 'Whalebone.'" Indeed, every day we find convincing proofs of the success of it, in such instances as 'Asteroid,' 'Audrey,' 'Miner,' and various others; whereas 'Cotherstone' formerly furnished an example of its success, being by 'Touchstone,' dam 'Emma,' by 'Whisker.' However, to end the subject as to 'Whalebone's' descendants, one could hardly do better than reply, as the late Mr. John Day used to, when interrogated by the inquisitive public as to which was the best of his Derby lot, " They are all good;" so, merely referring to the fact that 'Whalebone' himself won the

Derby in 1810, his own brother 'Whisker' in 1815, and that his descendants in the immediate line have won thirty-six Derbies, Oaks, and St. Legers, we may leave this illustrious family " alone in their glory."

The ' Sweetmeat' or ' Gladiator' blood appears to have caused quite a sensation, as well as called forth from the " authorities " upon such subjects their respective opinions, which appear to differ very considerably, the principal question at issue being as to the *staying* powers of his stock.

'Sweetmeat,' by 'Gladiator,' dam 'Lollypop,' by ' Starch,' or ' Voltaire.'—In order to arrive at a fair and proper conclusion as to a horse's merits, one would natur-ally suppose that they should be founded, not upon mere prejudice or imagination, but plain matter of fact; then, who that has ever made breeding and racing his study, taking for granted that he possesses a reasonable amount of the gift of discernment or understanding, will show that this horse was not only a combination of some of the best-proved blood in the world, and the *stoutest*, too, but also during his racing career shone forth a most brilliant star, having run twenty-four times, at two and three years old, and been but once defeated ; having met the best animals of the day, amongst others ' Alice Hawthorne,' for the Doncaster cup, which, one would imagine, ought to be a test of the staying qualities of a three-years-old, particularly when opposed by such an animal as the lengthy and everlasting stayer, 'Alice Haw-thorne,' the heroine of so many contests.

Then, with regard to his staying blood on his sire's side, we have had very many samples of its success ; such, for instance, as ' Blink Bonny,' whose dam was by ' Gladiator :' and we must also bear in mind, that if the ' Kingston' or ' Venison ' blood can stay, the fact of ' Gladiator ' being by the same sire, ' Partisan,' should prove no drawback ; nor should that of ' Glaucus,' being also by that horse, for we have most convincing proof of the staying powers of this blood in 'The Nob,' 'Nabob,' 'Trouncer,' and, though last not least, that magnificent specimen of " a horse," ' Nutbourne.' A certain doubt must exist as to whether ' Sweetmeat's' dam was by 'Voltaire' or 'Starch ;' if she were by the former, there can be little doubt that her relationship to ' Voltigeur ' ought not to deteriorate from her son's merits as a stayer. Be that as it may, upon the latter point one would imagine that the success of his stock in the Derby and Oaks, over the severe hills of Epsom, coupled with the fact that ' Dundee,' one of the stoutest horses of the day, even as a two-years-old, is the produce of a ' Sweetmeat' mare, ought to be powerful and convincing proofs as to the staying powers of his descendants : as instances of which we have ' Macaroni,' ' Mincemeat,' ' Mincepie,' 'Citron,' ' Sweetsauce,' ' Carbineer,' and many others. With regard to ' Sweetmeat's' shapes, his general outline was a subject for an artist, being symmetry all over, from his beautiful head to his tail, without a particle of lumber, but genuine bone, muscle, and sinew; if anything, leading one to the idea that he was slightly deficient in substance : his colour was a rich brown. My opinion is, that his dam was by ' Starch,' from my perfect recollection of the appearance of the latter animal when he stood at Walker's Horse Repository in Dublin, some thirty years ago, and, strange as it may appear, frequently in the company of a small brown bear : for, like ' Harkaway ' and his Newfoundland, 'Phryne ' and her goat, ' Kingston ' and his cat, ' Starch ' appeared very partial to Bruin's company.

In forming opinions as to the staying powers and other qualities of any horse's stock, it frequently happens that persons completely overlook the fact that the *dams* have *something* to say to the merits or demerits; yet people are inclined to forget it, and attribute the want of staying powers to the sire alone, totally overlooking the fact that the dam may have been a wretched, soft-hearted weed; or, even if a fine slashing mare, may inherit the "soft drop," from some strain totally different from that of the sire. It is likewise truly wonderful to what an extent people become prejudiced, and form hasty conclusions as to the distances certain animals can stay. How frequently some are led merely to confine the test of horses' qualities to the distance their respective blood or families may have " THE NAME" of being partial to; and thus keep them year after year trained for, and engaged in, stakes of a short course—simply because, at *two years old*, they have " *shown speed* "—overlooking the probability that trained on they would improve, &c. The consequence is, owners and trainers frequently follow like a " flock of sheep," and are carried away by prejudice, and never give the chance which their animals' *character* entitled them to, although their *reputation* denied.

With regard to the crosses which have appeared to suit this ' Gladiator ' or ' Sweetmeat ' blood, we have ' Dundee ' (by ' Lord of the Isles,' out of ' Marmalade,' by ' Sweetmeat '), whose extraordinary performances, as a two-years-old especially, ought to satisfy any reasonable person as to his staying powers, as well as that the ' Whalebone ' and ' Sweetmeat ' cross is desirable; other instances of which we find furnished in 'Mincepie' (winner of the Oaks) by 'Sweetmeat,' dam 'Foinnalla,' by 'Irish Birdcatcher;' 'Crater' by 'Orlando,' dam 'Vésuvienne,' by 'Gladiator;' 'Ledbury' (a remarkably nice horse, and a good one too) by 'The Cure' or 'Sweetmeat,' dam 'Themis,' by 'Touchstone;' and, though last not least, that first-class racehorse at *all distances,* 'Sweetsauce,' whose successes for the Stewards' and Goodwood cups, &c., beating large fields and the best horses of the day, in the commonest of canters, stamp him as a horse of *extraordinary* merit. His dam, the 'Irish Queen,' was by 'Harkaway,' grandson of 'Whisker.' 'Blink Bonny' furnishes an instance of where the 'Gladiator' and 'Melbourne' cross seemed to suit; and, strange to say, it has not been followed up.

There can be no question that in this blood, as in many others, there have been other successful alliances: but if the object be to test where it best suits, it follows that it must be by taking those cases where *first-class racehorses* have been the produce, not where a lot of moderate animals have "run a bit:" for instance, in the case of 'Sweetmeat,' we have proofs in favour of 'Touchstone' (which appears to suit with any other), 'Pantaloon' (which I believe a first-rate cross for any other), and the 'Whalebone,' whether through 'Irish Birdcatcher,' 'Touchstone,' or 'Economist.' It strikes me that a cross between 'Sweetmeat's' sons and 'King Tom' mares, or 'King Tom' and 'Sweetmeat's' daughters, would successfully rival any other. 'Citron' was, perhaps, for any distance, one of the best mares ever foaled (her dam, 'Echidna,' also dam of 'The Baron,' was by 'Economist'); and although reared by hand (her dam having died when foaling), was a first-class racehorse; and if she fail to produce one I shall be very much surprised, especially in the possession of her present owner, who never better displayed his judgment than when he purchased this mare—to my mind, worth as much money as any *untried* brood mare living.

The value of the ' Pantaloon' blood is undeniable, having furnished so many proofs, not alone as to speed (which I believe is their *forte*) and staying, but " *running strain :*" for, although some others occasionally produce one or two first-class animals, still few can compete with that in question as to number. Amongst others, in 1841, his son, ' Van Amburgh,' ran second to ' Coronation' for the Derby; and ' Satirist,' another of his sons, won the St. Leger, beating ' Coronation' and ' Van Amburgh,' his daughter ' Ghuznee' winning the Oaks in the same year; to which may be added ' Cardinal Puff,' ' Elthiron,' ' Miserrima,' ' Hobbie Noble,' ' The Reiver' (who ran second to ' West Australian' for the St. Leger), ' The Libel,' ' Hernandez' (winner of the 2000-guineas stakes), ' Legerdemain,' ' Clarissa,' ' Windhound,' &c.: the latter, there can hardly be a question, being sire of ' Thormanby;' ' The Libel,' grandsire of ' St. Albans,' &c.

With regard to the crosses with other blood (independently of those proofs in a direct line), ' Dundee' would appear to favour that between ' Pantaloon' and ' Sweetmeat,' ' Lord of the Isles,' dam ' Fair Helen,' being by this horse; as also ' Macaroni,' winner of the Derby, whose dam, ' Jocose,' is by him. Then, with regard to ' Thormanby' (supposing him to be by ' Windhound,' which I for one believe to be the case), ' Pantaloon' seems to mix successfully with the ' Mulcy Moloch' blood. My reasons for

believing that 'Thormanby' is by 'Windhound,' not by
'Melbourne,' are these: His colour (which is that of
'Pantaloon') — a contrast of his shapes (head and ears
especially) with those of 'Oulston' (also a son of 'Alice
Hawthorne'), by 'Melbourne' — the fact, that at the
time the question of the latter's impotency was very
much canvassed; and, finally, because, if I mistake
not, the mare was last served by 'Windhound:' which,
however, does not follow as a conclusive proof, as mares
frequently are with foal from first service. Moreover,
there is none of the 'Melbourne' plainness or appearance about 'Thormanby' in any respect. It seems
strange that 'Windhound' has not been better supported, for his family could all run; and I believe he
was considered by his owner and trainer no exception,
and would have proved so, had he not met with an accident. It appears he is now in the country, where stallions (like bacon-pigs) are plenty, and probably some of
them as fat likewise.

Another circumstance worthy of remark, as regards
the mixture of the 'Pantaloon' blood with that of
'Touchstone' (of which the Marquis of Westminster
thought so highly), is, that 'Alice Hawthorne' produced
by 'Touchstone' a horse called 'Findon' (whose pedigree would have led me to expect wonders); yet he
never distinguished himself on the course, and probably
will never get the *chance* of doing so at *stud*, although he
might succeed, for "blood will tell." Then we find
'Thormanby,' out of the same mare, by a son of 'Pantaloon,' dam 'Phryne,' by 'Touchstone,' showing the value
of the " mixture."

Before closing my remarks upon this " running family"—instances, his own brothers, ' Elthiron,' ' The Reiver,' and ' Hobbie Noble'—I must express my surprise at ' Hobbie's' not having, ere now, more highly distinguished himself; my belief being, that even now, like ' Surplice' and others, there must be some extraordinary cause : for there are frequently many, which it is not my place or intention here to allude to. However, there is one patent fact, viz. that he was really a good and a first-class two-years-old; and a much greater proof could hardly be, than that 6000 guineas were paid for him to win the Derby : a feat which, however, he failed to accomplish, although every precaution was taken, a guard of honour having accompanied him to Epsom, reminding one of the history of Caligula and Incitatus.

' Leamington's' superior qualities furnish proof of the excellence of the cross between ' Pantaloon' and ' Sir Hercules,' being by ' Faugh-a-Ballagh,' dam by ' Pantaloon,' as also ' The Marquis' (winner of the 2000-guineas stakes and St. Leger), whose dam, ' Cinizelli,' is by ' Touchstone,' dam ' Brocade,' by 'Pantaloon.' The dams of 'Toxophilite' and the ' Prime Minister' are also by him, as well as the dam of ' Young Melbourne' (sire of 'General Peel'), as also 'Emily,' the dam of ' Irish Queen;' showing that this blood appears to suit with various strains, but more especially with ' Whalebone.' To my mind, ' Pantaloon' blood cannot be excelled.

The ' Voltaire' blood appears to have formed one of the subjects of recent discussion, especially as regards the merits of his son ' Voltigeur.' In these days of warfare, the fanciers of horseflesh appear to have their differences to solve, from the manner in which the war has been carried on by the partisans of this blood, and that of ' Gladiator' or 'Sweetmeat;' the respective champions put forth by each being 'Voltigeur' and 'Sweetmeat.' If the quality of staying be so great a desideratum (which doubtless it is), surely 'Voltaire' and his stock have furnished ample proof of their excellence in this respect; instances of which we find in the following:—'Charles the Twelfth,' winner of the St. Leger after a dead heat with 'Euclid,' also winner of the Doncaster and Goodwood cups twice; 'Voltigeur,' winner of the Derby and St. Leger (curiously enough, also, after a dead heat with 'Rusborough' for the latter, a son of 'Tearaway,' by 'Voltaire'). 'Tearaway,' who proved himself one of the very best horses ever foaled, under enormous weights, and for any distance; and sire of 'Kingstown,' who ran second to 'Wild Dayrell' for the Derby; 'Semiseria,' whose performances were first-class, having beaten 'Nutwith,' 'The Cure,' and many others; 'Buckstone,' by 'Voltigeur;' 'Cavendish' (a remarkably good, and good-looking one); 'Skirmisher,' winner of the gold cup at Ascot, and various other races; 'Hartington,' winner of the Cæsarewitch, 1862, &c.; not forgetting my *especial favourite*, that superior racehorse 'Vedette,' an animal upon which I purpose making a few remarks at the close of my observations upon my selection from the sires of the present day.

The descendants of 'Voltaire' are in general what may be best described as fine, slashing specimens of the racehorse, with plenty of length and racing shapes, being invariably of a rich dark-brown colour, with fine freedom of action and propelling power. Their staying qualities can hardly for one moment be questioned; and as to the idea entertained by some that they are deficient in speed, with great deference to the opinions of such parties I must take the liberty of differing with them, founding my right to do so on simple matters of fact, and on reference to the racing records, which so repeatedly and distinctly bear testimony to the truth of my statement: for one fact appears to escape the recollection of so many persons —that because certain horses happen to possess staying qualities, it does not therefore follow that they *must* be *deficient* in *speed*, although it is quite true that there are many "half-mile squibs" that cannot stay one yard beyond their distance.

With reference to the strains of blood which seem to cross successfully with the one in question, it would appear that the 'Whalebone' (which appears to "bend" well with any) has best suited: instance 'Vedette' by 'Voltigeur,' dam by 'Irish Birdcatcher;' 'Tearaway' by 'Voltaire,' dam 'Taglioni' by 'Whisker;' 'Rusborough' by 'Tearaway,' dam 'Cruiskeen' by 'Sir Hercules;' 'Cavendish,' 'Hartington,' and 'Buckstone's' dams by 'Touchstone,' 'Zetland' by 'Voltigeur,' dam 'Merry Bird,' by 'Irish Birdcatcher,' besides many others. No doubt it would be desirable to cross with speedy mares, such as 'Birdcatcher.' An addition in proof of the staying qualities of this blood is found in the fact, that 'War Eagle's' dam, 'Valentine,' was by 'Voltaire;' which horse's merits will be referred to under the head of 'Lanercost;' 'Piccaroon,' grandsire of 'Old Calabar,' &c.

'Lanercost,' by 'Liverpool,' dam 'Otis,' by 'Bustard,' was no doubt a great horse; remarkable for his weight-carrying, staying powers, which he has so well transmitted to posterity, and which do not appear to be losing any of the *prestige* which is attached to his blood. As a racehorse he was of the first class, and as a proof of the estimation in which his merits were held he was handicapped to give to the renowned 'Alice Hawthorne' 51 lbs. when she was a four-years-old : a nice undertaking, no doubt, to accomplish !    Amongst his descendants we find 'Van Tromp,' winner of the St. Leger, Doncaster, and Goodwood cups, beating the best horses of his day; 'War Eagle,' winner of the Doncaster cup, beating the 'Hero;' he ran second to 'Cossack' for the Derby, second to 'Peep-o'-Day Boy' for the Chester cup, giving him 20 lbs. at the same age; and second to 'The Widow,' aged for the Cambridgeshire, giving her 17 lbs., beating thirty-four horses, to all of whom he gave weights —to several of his own age as much as three stone, being the 'Yellow Jack' of previous days.   'Catherine Hayes' (his daughter), winner of the Oaks, 1853, proved herself otherwise an exceedingly good animal; 'Ellerdale' (another of his daughters), a very good mare, dam of 'Ellington' (winner of the Derby); 'Gildermire' (second for the Oaks after a dead-heat with 'Governess'), 'Wardermarske,' 'Summerside' (winner of the Oaks), 'Ellermire,' &c.   Then we find 'Ivan' (by 'Van Tromp'), that probably, as a racehorse, was about the best of his day, of which his running with 'Vindex' and others bears testimony; then come 'Van Galen,' sire of 'Tim Whiffler;' 'Union Jack,' by 'Ivan;' also 'Rapparee,' whose dam, 'Lady Alicia,' is by 'Lancercost;' 'Colsterdale,' own brother to 'Ellerdale.'   As to the crosses which appear to suit this blood, we find 'Union Jack' bears proof in favour of that with the 'Sir Hercules' strain, through 'Caprice,' whose sire was 'Coronation;' 'Rapparee,' by 'Rataplan,' dam 'Lady Alicia,' by 'Lancercost.'   Then, in favour of that with 'Melbourne,' we find 'Fairwater,' by 'Loup-

Garou' (a son of 'Lancrcost'), dam 'The Bloomer,' by 'Melbourne,' half-sister to 'Ely,' both the property of that thorough sportsman Mr. Cartwright.

'Ion' is now represented by 'Wild Dayrell,' 'Tadmor,' 'Pelion,' 'Buccaneer,' and 'Horror.' It is hardly necessary to state, that the most fashionable and most deserving of patronage is the first-named. For size, racing shapes, and, in point of fact, for every other qualification that can be desired in the racehorse, 'Wild Dayrell' is unsurpassed by any animal living. There can be no question as to his having had a very fair chance up to the present time, and it is equally true that his stock have to a great extent upheld the renown of their sire; amongst which we find 'Buccaneer,' 'Wild Agnes,' 'Avalanche,' 'Hurricane,' 'Horror,' 'Investment,' 'Dusk,' 'Wildman,' and others. Further remarks upon this animal will be found in a special article hereafter. 'Pelion' by 'Ion,' dam 'Ma-Mie,' by 'Jerry,' was perhaps one of the best (if not *the* best) mile-horses of his day, and is worthy of far more patronage than he appears to have received, being an exceedingly good-looking horse, of a beautiful dark-brown colour, and in other respects possessing those qualifications which ought to have induced the public to have supported him for their own sake. Full well I remember, without reference to any racing calendar, his running at Chester, when three years old, where I happened to have a mare named 'The Deformed,' engaged in the same race, which was won by 'Exact.' Few *really* knew the superiority of these three animals. For my part I cannot understand why 'Pelion' has not received more patronage, and confess a great prejudice in his favour; and, without entering into a rigmarole dissection of his shapes, pronounce him, in my humble opinion, one

of the best-looking horses in the kingdom. It would take
a great deal to persuade me, that if 'Pelion' had had as
good a chance as others he would not have distinguished
himself quite as highly as many other sires, that are
thought more highly of; even with the limited chance he
has received he has produced animals that could run.
Although they have been confined to short distances they
have been, like many others, condemned as milers or half-
milers, because they have never been half trained or tried
the distance: instances of which I have so frequently witnessed. His own brother 'Poodle,' and others, could stay.
My remarks upon 'Buccaneer' will be found hereafter.

With regard to the most successful crosses with the
'Ion' blood, one patent fact presents itself, viz. that the
'Bay Middleton' or 'Sultan' suits—of which Wild Day-
rell' bears a striking proof.

'Bay Middleton' by 'Sultan;' dam 'Cobweb,' by
'Phantom,' 'Sultan' by 'Selim.'— Competent judges
appear to differ very much in their estimation of the
qualities, perfections, and imperfections of this blood;
some maintaining that 'Bay Middleton's' descendants are
leggy, tall, overgrown, and weak, and in many instances
roarers, their only *forte* being speed; others differing
materially in their opinions as to their merits.

It can hardly be denied that he had a very great
chance of distinguishing himself amongst others afforded
by his late noble owner, who purchased him for the large
price of 4000 guineas as a sire, and also a very large stud
of brood mares for the purpose of breeding from him; yet
the records of racing cannot furnish any great evidence of
his success as a sire; on the contrary, like a great number
of other really first-class racehorses, he was a comparative

failure: for with the exception of 'The Flying Dutchman,' 'Andover,' and 'The Hermit,' none of his produce proved first-class ('Vanderdecken,' own brother to 'The Dutchman,' being a very weak-leggy and indifferent sample), whereas very many were really worthless. He was himself a magnificent specimen of the racehorse; according to the opinion of most experienced judges, about the best horse that has appeared in the present century: be that as it may, it by no means follows that his sons and daughters may not prove most valuable and successful at stud. As to the racing merits of his stock, a few instances are furnished in 'Flying Dutchman,' 'Andover,' 'Autocrat,' and 'Fly-by-Night' (a much better animal than generally supposed). Then as to stud, 'Ellen Middleton' (dam of 'Wild Dayrell'), 'Ennui' (dam of 'Saunterer,' and grandam of 'Liddington'), 'Blister' (dam of 'Mainstone'), 'Bridal' (dam of 'Special License,' a long runner), 'Sunflower' (dam of 'Sunbeam'), 'Rose of Cashmere' (dam of 'Wild Rose'), were all his daughters. 'Haggish' (grandam of 'The Hadji'), was also by him. 'Pocahontas' (dam of 'Stockwell,' &c.), was by 'Glencoe' (son of 'Sultan') and 'Mainbrace' (dam of 'Fisherman'), was out of a 'Bay Middleton' mare. The grandams of 'Scottish Chief,' 'Rupee,' and 'Stampedo,' were by 'Bay Middleton.' In my opinion, his sons and daughters will prove the fact that, as in many similar cases, they fail to prove racehorses; yet they inherit and hand down their superiority to subsequent generations: such, for instance, as in the case of 'Pylades' (son of 'Surplice'), sire of 'North Lincoln,' an extraordinarily good animal for his distance. Mr. Goodwin of Hampton Court, whose great experience must entitle his opinion to the highest respect, says, " He

o

believes ' Bay Middleton ' was a long way the best horse
we have seen in this century," and adds,

> " ' He was a horse; take him for all and all,
>     We shall not look upon his like again.' " *

Amongst the various strains of blood, few will be
found with more adherents than these; and justly so,
for reference to " The Stud-Book," and " The Racing
Calendar," must convince any impartial reader that they
have proved most successful. ' Melbourne' especially
has proved most valuable, being sire of ' West Australian,' winner of the double event, Derby and St. Leger,
and one of those animals justly entitled to compete, with
a few others, for the title of " Champion," as the best
horse of modern days. ' Melbourne' was also sire of that
magnificent mare ' Canezou '—to my mind the " finest,"
if not the " best," we have seen for many years : for,
although defeated by a head by ' Surplice' for the St.
Leger, had she not lost a plate at the " Red House," the
result might have been different. Be that as it may, her
other performances (under enormous weights) proved her
a wonderful mare; and, to my mind, we have not seen a
grander specimen of one.

' Melbourne ' being descended from ' Comus,' entitles
him, to a certain extent, to claim credit for the performances of ' Hetman Platoff ' (the latter's grandson) : as
evidence of the value of this blood, ' Cossack,' by ' Hetman Platoff,' won the Derby. The dams of ' Daniel

* Query—' West Australian,' ' Stockwell,' ' Flying Dutchman,' ' Vedette,' ' Faugh-a-Ballagh,' ' Harkaway,' ' Melbourne,'
(by ' Humphrey Clinker,' dam by ' Cervantes,') and ' Hetman
Platoff,' by ' Brutandorf,' ' Don John's' dam by ' Comus.'

O'Rourke,' winner of the Derby, and of 'Knight of St. George,' winner of the St. Leger, were also by him. 'Gamester,' by 'Cossack,' won the St. Leger. 'Special License' was also by him. 'Springy Jack,' second for the Derby (to 'Surplice'); 'John Cosser,' and 'Nea-sham,' winners of the Northumberland plate, were also by 'Hetman Platoff.'

The value of the 'Melbourne' strain has been wonderfully exemplified in later days by the following proofs :—His daughter, 'Blink Bonny,' winner of Derby and Oaks; 'Thormanby,' by 'Melbourne' or 'Wind-hound;' 'The Wizard,' winner of the 2000-guineas stakes, and second in the Derby, by 'West Australian;' 'Blair Athol,' winner of the Derby and St. Leger, dam 'Blink Bonny;' 'Cymba' (winner of the Oaks, 1848), 'Marchioness' (Oaks, 1855), both by 'Melbourne;' and 'Summerside' (Oaks, 1859), by 'West Australian;' 'Sir Tatton Sykes' (winner of the St. Leger, 1846), by 'Melbourne.' The dams of 'Lord Clifden' (St. Leger, 1863), 'The Slane' and 'Limosina' ('Charity'), were also by him. In addition to the above there are innumerable proofs of the value of this blood : amongst others, 'Hulston,' son of 'Alice Hawthorne,' besides Lord Glasgow's promising 'Young Melbourne,' sire of 'General Peel,' winner of the 2000 guineas, and second in the Derby and St. Leger; 'Rapid Rhone,' &c.

'Melbourne's' sons and daughters are remarkable for their great size, substance, and soundness, large bone, wide hips, and immense power, but racing-like, with plenty of length; their *contour* is plain, their heads invariably large, plain, clean, and bony, and their ears frequently 'lopped.' Judging of things as they stand in

the present day, the blood of ' Melbourne' ranks second
to none.

As to the crosses which appear to have best suited with
' Melbourne,' we have in the ' Whalebone' line ' Blair
Athol,' ' West Australian' (both very much alike as to
breeding), ' Lord Clifden,' ' Fazzoletto,' ' Stockade,' ' Limosina,' &c. ; and in that of ' Gladiator ' or ' Sweetmeat,'
' Blink Bonny,' &c. : a strange fact being that the latter
cross has been almost totally disregarded, although the
symmetry of the ' Sweetmeats' alone would suit the
other strain, ' The Nob,' by ' Glaucus,' dam ' Octave,' by
' Emilius,' grandam ' Whizgig,' by ' Rubens.'

While estimating the value of the numerous strains of
blood, it would, indeed, be an oversight to omit referring
to this one, especially when it can hardly be denied that
stayers are " few and far between." For size, shape, power,
and endurance, it is very questionable if there rank any
preferable. It is an admitted fact, that " good big ones
will beat good little ones ;" and it is remarkable that all
the stock of this horse are of fine size ; and when we look
at ' Nutbourne,' as one instance, probably it would be
very difficult to find in Europe a more magnificent sample of the powerful thoroughbred, particularly when it is
remembered that even at two years old, an age at which
such large animals do not display their real *forte*, or
powers, and at a distance more suited to early light-framed
horses, he cut down the very best of his year, and had he
not met with a *contretemps* in the Derby, was doubtless the
most dangerous opponent to the winner ('Thormanby'),
although the course was anything but suited to a horse of
his heavy frame, especially that part of it where he met
the accident ; which was not, however, one that might be

termed natural, or one arising from any infirmity in his legs, but in reality from having jumped across the road after descending that trying hill.

'Rupee' was by 'The Nabob.' We have also had proofs of the quality of this horse's descendants. Even in the present year, 'Vermuth,' winner of the grand prize at Paris, defeating the winners of the English Derby and Oaks, 'Blair Athol' and 'Fille de l'Air;' besides, 'Bois-Roussel' (winner of the French Derby) is by 'The Nabob,' son of 'The Nob;' as was 'Trouncer,' an extremely good horse at all distances, and under heavy weights, to carry which he was so well formed.

In my opinion it is very questionable if there be at this moment any breed more desirable, especially for the improvement of the horse in a general point of view, than the one in question; for it must be borne in mind that this strain has been sadly neglected and unpatronised, while the wholesale run has been upon others. Seldom have French connoisseurs better displayed their judgment than when they purchased 'The Nabob;' and yet the British public seem surprised that the principal prizes should be snatched from them by horses whose ancestors have been purchased regardless of price, and offered at figures to the public service which must encourage breeders. The natural result will doubtless be, that ere long the French division will be generously and gratefully offering an "allowance" to English-bred horses, in return for the courtesy and leniency which has for years been extended to them. For with a better climate, as good provender, and as good trainers, it appears strange that they should be held in less esteem than British-bred horses.

'Weatherbit,' by 'Sheet Anchor,' dam 'Miss Letty,'

by 'Priam.'—This strain of blood appears to be very much fancied, by certain Northern breeders especially. No doubt he has handed down some very valuable samples of his quality; the best of which appear to be 'Beadsman,' 'Appenine,' 'Prince Arthur,' and 'Neptunus,' as race-horses. 'Sheet Anchor,' his sire, has done good service at the stud; 'Mainbrace,' dam of 'Fisherman,' being his daughter; as also 'Yard-Arm,' dam of those sound and good brothers, 'Gunboat' and 'Lifeboat'—a class of animal not to be found every day: being a great contrast, in a general point of view, to the very weedy, light-boned horses, with which the country is so overrun: besides 'Netherton Maid,' grandam of 'Big Ben,' and 'Skiff,' dam of 'Cymba' (Oaks, 1848). This blood appears to have hit well with that everlasting and elastic 'Whale-bone,' through the animals above named: the dam of 'Beadsman' being by 'Touchstone;' and the "two bro-thers" being by 'Sir Hercules.' The fact of 'Weatherbit' having the 'Priam' strain in his veins, should be a further recommendation. 'Cymba' is a proof of the success of the cross with 'Melbourne.'

In addition to the various strains of blood which have been referred to, there are some others which have of late years being gradually dying out: 'Priam' especially; al-though 'Surplice,' 'Beadsman,' and 'Chevalier d'Indus-trie' retain it. The two latter are, no doubt, not on so large a scale as many of the sires of the present day; still they are nice, wiry, racing-like horses, and are most likely to prove successful at the stud, especially if crossed with good-sized mares. But the very great competition in the present day renders it almost an impossibility that all the sires deserving of support can receive it: for while that prejudice in favour of the double cross of the 'Whalebone,' "the ready-money cross," exists, breeders are hardly likely to increase the risk which attends such speculations: therefore, as a natural consequence, many valuable sires are likely to pass away, without even a moderate chance; for where, according to the records of each at present at stud, there are not more, on an average, than about six mares to one sire; in addition to which fact, that "the run" is all upon a chosen few, and, as a matter of course, the picked mares are sent to those horses, — how can all have a fair chance?

As an instance of how frequently valuable sires are passed over, that promising young one, 'Ivan,' imported to Ireland by the late Mr. Courtney, would in all probability have never had a chance, had he not produced 'Union Jack;' and would have but taken old 'Sir Hercule's' post — "improving the half-bred stock in Ireland." It has been stated by many, that "in-and-in" breeding tends to diminish, not only the stamina and powers of endurance, but the size also.   Every day furnishes proofs to the contrary; at least so far as "the double mixture" of 'Whalebone.'   Take 'Asteroid,' 'Big Ben,' 'Audrey,' 'The Marquis,' and, of more recent date, 'The Miner;' besides numerous others.   Still, there are innumerable proofs where that incessant cross between 'Stockwell' and 'Touchstone,' and similar mixtures of the same strains, have merely turned out "short runners."   For instance, 'Exchequer,' 'St. Alexis,' and that splendid mare 'Emily,' and a host of others: whereas when crossed with other blood, such as 'Lancercost' ('Caller-Ou,' for instance), and other staying strains, they have invariably proved stout.

Having glanced over the various sires, whose descendants seem in the present day to occupy the attention of breeders and purchasers, I shall proceed to lay
before the reader a few remarks upon a chosen few, selected from the vast number at present at the public service; admitting that there may be, and are perhaps others,
equally deserving of patronage. However, every person
is entitled to have his own opinion; and the following
comprise those which, to my mind, are best calculated to
benefit the breeders or purchasers of thoroughbred stock.
From the number of horses bred annually, it would be
idle to suppose that there must not, as a necessary consequence, be many deserving of notice; for, as the adage
goes, "there are as good fish in the sea as ever were
caught." However, as there are at present about three
hundred advertised, I shall proceed to select and recommend the following: forming my conclusions as to their
merits on the grounds of their general recommendation,
taking into consideration breeding, size, shape, substance,
and performances, as well as the other qualifications requisite in the racehorse, and more especially "running
families."

### 'AMSTERDAM.'

A bright bay horse, nine years old, by 'The Flying
Dutchman;' dam 'Urania,' by 'Idleboy,' by 'Satan;'
her dam, 'Venus,' by 'Langar,' out of 'Vesta,' by
'Governor.'

'Amsterdam' is remarkable for his symmetry, wonderful length, and racing shapes otherwise; his hind-quarters
being beautifully formed and placed. If speed be the
great desideratum in a racehorse, he certainly was gifted

with it to an extraordinary degree; for it is very questionable if (especially at five years old) there was an animal in training that could have beaten him, at even weights, one mile : in proof of which, any dubious reader has but to refer to the records of racing to satisfy himself of the fact, which is there most distinctly recorded, that carrying the top-weight, exceeding, in many instances, nine stone, he beat and gave weight to the very best horses of the day, including 'Twilight' (to whom he conceded 6 lbs.), 'Zuyder Zee,' 'Starke,' 'King of the Forest,' 'Lady Trespass,' 'Atherstone,' 'Crater,' 'Comforter,' 'Prelude,' 'Libellous,' 'Lord Berkeley,' and many others.

His outline and contour, at a glance, display the true shapes of the racehorse, and if he possessed a slight shade more substance he would be perfection : still he is of that steely, wiry frame, totally void of a particle of lumber or coarseness, which sometimes makes an animal at first sight appear deficient in substance. There can be hardly a doubt that, with a fair chance, which he is likely to get in his present harem, he will prove the sire of nice stock ; especially if crossed with mares of substance : which sometimes require a horse of this stamp to fine down coarseness, as well as improve deficiency of length.

### 'ARTILLERY.'

By 'Touchstone;' dam 'Jannette,' by 'Irish Birdcatcher;' grandam 'Perdita,' by 'Langar.' A bay horse, with the usual white marks of his sire's stock, and good size. He was a racehorse, although not so fortunate as might have been expected from his blood (which is a combination of the best strains), as well as having good looks.

Having been hired to run out his engagements in the Derby, &c. his career was not a brilliant one; still, he ranks far before many others as worthy of patronage at stud, that are, in reality, mere "squibs," when compared to him in any respect; although supported by breeders merely on account of the *prestige* of victory, which does not at all times test the true merits of either men or horses.

'Artillery' can hardly fail, with a fair chance, to get racehorses. He bears a striking resemblance to 'Mountain Deer' (also by 'Touchstone') in many respects; although, perhaps, not so lengthy or powerful in his hind-quarters and thighs.

Exception has been taken to him by some, in regard to his being rather flat or light in his back-ribs: but that is a peculiarity to most of the 'Touchstones,' and by no means so objectionable in the racehorse, especially when counterbalanced by fine loins, &c., which he possesses; being like almost all the breed, both on the side of sire and dam, remarkably good behind the saddle. It was also thought, that when about to contend for some of his engagements he appeared as if he had exhausted all his " ammunition;" which partly left the impression of the deficiency referred to.

If he do not share the fate of 'Irish Birdcatcher' until the close of his career (in whose box he probably now stands), he cannot fail to get racehorses.

### 'BEADSMAN.'

Nine years old: a brown horse, by 'Weatherbit;' dam ' Mendicant,' by 'Touchstone;' grandam 'Lady Moore Carew,' by 'Tramp;' great grandam 'Kite,' by 'Bustard;' 'Weatherbit,' by 'Sheet Anchor;' dam 'Miss Letty' (winner of the Oaks), by ' Priam,' ' Orville,' ' Buzzard.'

This horse must rank amongst my chosen few, not merely because he is recorded amongst the winners of the Derby, but from other qualifications, which must entitle him to the highest consideration. He is a model of a racehorse, without a particle of lumber; a wiry, neat animal, his colour of the richest brown, and his blood for every requisite, speed, stoutness, "running family," both on the sides of sire and dam, such as cannot be surpassed : for, through his sire, he inherits the immediate blood of ' Priam' (rather deserted of late), while that of his dam requires no comment, nor do her performances as a racehorse. If one could wish to add in any respect to the many qualities which must recommend this horse to the breeder, it might be a little more size, if the object were to breed for general purposes; yet, although there are others on a larger scale, his superior is hardly amongst the untried division : and I shall be very much surprised if ' Beadsman' does not turn out as successful at stud as he did during his racing career.

' Weatherbit' (his sire) being so much fancied by certain Northern breeders of judgment and experience, it is unnecessary to remark that this horse's dam being ' Mendicant,' will at least not lessen his value in their estimation, or render him less deserving or likely to take his sire's place at a future day. Taking into consideration that the success of animals for the Oaks must bear great testimony in favour of their quality, it should not be forgotten that the following mares by ' Priam' won that race,—' Miss Letty,' in 1837; ' Industry,' in 1838; ' Crucifix' and ' Welfare,' first and second, in 1840.

### 'Big Ben.'

Six years old, by 'Ethelbert;' dam 'Phœbe,' by 'Touchstone;' grandam 'Netherton Maid,' by 'Sheet Anchor,' 'Tantivy,' 'Myrtilla.'

A fine horse, and, as far as size, most appropriately named, being an animal of great power, and his blood undeniable—a combination of the winning strains of the day. His performances were very good, which will appear upon reference to the racing records, beating large fields, comprising such animals as 'Dundee,' 'Fairwater,' 'Dictator,' 'Walloon,' 'Folkestone,' and others, highly tried and fancied by their owners. His fine size, substance, shapes, and breeding, must highly recommend him for any purpose, as he is one of the few stallions of the present day possessing that extra power and size so very desirable, and so seldom found.

### 'Buccaneer.'

A dark-bay horse, seven years old, by 'Wild Dayrell;' dam 'The Little Red Rover' mare (also the dam of the well-known 'Cruiser').

A great number of animals come fairly under the denomination of racehorses, but there are others that, if it were possible to find a higher term, are justly entitled to it; and pre-eminent amongst these stands forth 'Buccaneer.' His performances as a racehorse were first-class; and but for an accident might, and doubtless would, have been more distinguished. The field of horses which he beat at Ascot, including such as 'Prétendant,' 'Cosmopolite,' 'Lava,' 'Amsterdam,' 'Fravola,' 'Mainstone,' 'Nutbush,' 'Gabrielle d'Estrées,' 'Elcho,' besides many

others, independently of his other victories, must place him
in the first rank against the most dangerous opponents,
where I have little doubt he will at stud, as at racing,
take his part as successfully as his namesakes always did.
He has plenty of length, substance, and racing shapes;
which I do not pretend to dissect, as I merely write
from recollection of the animal, only adding that, in my
opinion he will prove at stud a worthy son of his mag
nificent sire.

### ' CARACTACUS.'

A bay horse, five years old, by ' Kingston;' dam
' Defenceless,' by ' Defence;' her dam by ' Cain,' out of
' Ridotto,' by ' Reveller.'

This horse appears now to be at least the most fashionable, as well as the most desirable representative of the
' Venison' blood; the spirited proprietor of the great
Middle Park stud having some time ago experienced a
loss by the death of his favourite, ' Kingston,' the seri-
ousness of which has become subsequently more apparent through the successes of his sons and daughters,
' Queen Bertha,' winner of the Oaks, &c. A very curious
prejudice seemed at first to exist in the minds of many
that the ' Kingstons' could not stay, simply because two
or three of them showed extraordinary speed, yet deficiency in the other power; but that circumstance was
attributable to certain causes, which are frequently overlooked, or never understood by many, viz. that the imperfections or failings of dams must have a certain
influence, whether arising from natural or constitutional
causes, as well as the fact that horses hammered about
as he was, running the most severely-contested and

longest distances, most gamely, and always to form, as Goodwood cups and Northumberland plates bear testimony of; such horses can hardly be expected to be in their proper stud form for some time, after having been dried up for years, in racing condition.

'Kingston' was, in every respect, bred to stay; his sire, 'Venison,' proved himself a horse of undeniable stoutness; and it is worthy of remark, that during his career as a three-years-old he travelled on foot his circuit, which the more modern wonders, 'Fisherman' and 'Rataplan,' accomplished by rail; as it is stated upon reliable authority that he walked upwards of nine hundren miles, ran fourteen, and won twelve times. So much for 'Venison's' season! It seems strange that a belief should exist that the 'Venisons' display a sort of reciprocity of taste as to fondness of human flesh: probably from recollections of 'Cruiser,' 'Cariboo,' and 'Vatican.' A circumstance worth relating happened, with regard to 'Kingston,' when about to run for the Derby; all sorts of rumours being in circulation, amongst others that he was a "man-eater!" An acquaintance of mine, who had backed him, and who professes to be a judge of racehorses (and one who has a wonderful flow of the phraseology), became alarmed; having been informed by some "wiseacres" that such was the fact, he took care to have a look after the horse at exercise (as well as his money), and having repaired to the necessary quarters for the purpose, on his return informed me that he found him, after his usual work, walking as quietly as a lamb beside his boy, with his head almost resting on the lad's shoulder. Of the docility of 'Kingston' I have been a frequent witness, when he was in training and at stud. As to his son 'Caractacus,' he is in every way most worthy the notice of breeders, who fancy the 'Venison' and 'Defence' blood, (and who could object, to the latter especially?) more particularly as he, as well as bearing a marked resemblance to his handsome sire in other respects, has a good temper, which the other was possessed of to perfection; proving the fact, that good and kind treatment has its effects with such animals, as with most others.

'Caractacus's' performances were—independently of his winning the Derby—good; his success for the latter being a surprise, no doubt, to many, and stamping him as a very superior racehorse. He is not one of the large stamp; but, like his sire, a nice-sized, level-made horse, with plenty of quality and racing points, and most likely to get racehorses.

### 'CAVENDISH.'

A dark-bay horse, eight years old, by 'Voltigeur;' dam 'The Countess of Burlington,' by 'Touchstone;' her dam 'Lady Emily,' by 'Mulcy Moloch,' out of 'Caroline,' by 'Whisker.'

This horse is, and at all times was, an especial favourite of mine. I have a perfect recollection of him just previous to his first race at York (which he won very easily), when he struck me as being an extraordinarily good-looking two-years-old. His blood is undeniable, and it will surprise me if he does not prove a successful sire, although his career on the turf was of short duration; from what cause I know not, but most probably one of those accidents to which horses are so liable, however well formed or sound they may be by nature. His combination of blood is first-class, especially for staying qualities; in proof of which we have his own brother 'Hartington' (winner of the Cæsarewitch), and numerous other instances : besides, upon reference to the blood of 'Vedette,' they would appear almost full brothers in blood, both being by 'Voltigeur,'—the one out of a 'Touchstone,' the other an 'Irish Birdcatcher' mare ; and if there were any drawback on the side of either as to the grandam's pedigree, it certainly is not on that of 'Cavendish.' His own performance as a two-years-old proved his speed ; and as to his general contour, he was as good-looking a two-years-old as one could wish to see.

### 'Chevalier d'Industrie.'

A chestnut horse, ten years old, by 'Orlando ;' dam 'Industry,' by 'Priam ;' grandam 'Arachne,' by 'Filho da Puta,' 'Treasure,' by 'Camillus.'

This is a remarkably nice horse; the superiority of his blood is unquestionable, being a combination of the very best, and more especially as his dam ('Industry,' winner of the Oaks) was by 'Priam.' He proved himself a racehorse, and is a very wiry, level, lengthy animal, without lumber. He is closely allied in blood to 'Surplice,' but is of a totally different stamp to the latter in many respects ; 'Surplice' being upon a much larger scale, although perhaps not possessing, in various points, the racing shapes of 'The Chevalier,' who is a nice, neat specimen of the racehorse. His ancestors on his sire's side, as also on that of the sire of the dam, have all for a number of years, with the exception of 'Camel,' won the Derby or St. Leger ; viz. 'Orlando' (Derby, 1844),

'Touchstone' (St. Leger, 1834), by 'Camel,' by 'Whale-bone' (Derby, 1810), by 'Waxy' (Derby, 1793), 'Industry,' by 'Priam' (Derby, 1830), by 'Emilius' (Derby, 1823), by 'Orville' (St. Leger, 1802), by 'Beningbrough' (St. Leger, 1794).

### 'CLARET.'

A brown horse, twelve years old, by 'Touchstone;' dam 'Mountain Sylph,' by 'Belshazzar,' out of 'Stays;' own sister to 'Incognita,' by 'Whalebone.' He is own brother to 'Mountain Deer,' 'Champagne,' and 'Sylphine;' all good animals, and descended from the most running strains to be found in the records of racing: he was himself a very fair horse; his brother 'Mountain Deer,' and his sister 'Sylphine,' ranking in the first class. To my mind, 'Mountain Deer' is about the best-looking horse I ever beheld, and is a very great loss to the country; as he had proved himself (considering the chances he had) more successful than many sires that have had better opportunities, and are of longer standing; and my conviction is, that he would have replaced his sire in due time, had he remained in England. Claret is free from white (which some dislike), being of a rich, dark-brown colour, has already produced some winners, and with a fair chance is likely to supply more.

### 'CRATER.'

A bay horse, seven years old, by 'Orlando;' dam Vésuvienne,' by 'Gladiator;' her dam 'Venus,' by 'Sir Hercules,' out of 'Echo,' by 'Emilius.' 'Crater' proved himself, no doubt, a superior racehorse, beating very large

P

fields; for instance, in the Hunt cup at Ascot, wherein he defeated horses of all ages, giving away a great deal of weight, although his performances were principally confined to a mile; still, although only third in a large field for the Amport stakes at Stockbridge, it was hardly to be expected he could give 20 lbs. to such a horse as 'Northern Light,' who was the same form as 'Cape Flyaway;' nor yet the year and 7 lbs. to 'Ariadne.' In shape, size, &c., he is good-looking enough for anything, and he struck me at all times as being an exceedingly true and good-tempered animal: both horse and trainer furnishing perfect specimens of knowing their business. As to his pedigree it equals any on record, being of the same mixture of blood as 'Dundee;' there are few (if any) of the untried division more likely to produce racehorses.

## 'De Clare.'

A bay horse, twelve years old, by 'Touchstone;' dam 'Miss Bone,' by 'Catton;' 'Franby's' dam by 'Orville.'

This horse's breeding is undeniable; as to shape he is a fine specimen of the racehorse, and was an undoubted good one, although unfortunate, having been, up to within a few hours of the race for the Derby, first favourite at a ridiculously short price, but meeting with an accident he did not even start—a sort of fatality appearing to prevent the noble owner gaining the prize bearing his name. 'De Clare' is a horse of great power and length, with plenty of bone and substance; and coming from such running strains, not alone on his sire's side, but through his dam (likewise the dam of such first-class horses as 'Longbow,' 'Boiardo,' &c.), with a fair chance, he can hardly fail to prove successful at the stud. Exception has been taken to the formation of his shoulders, which, however, is a peculiarity in many of the 'Touchstones.' Had this horse been recorded amongst the winners of the Derby, he would have received more patronage as a sire, as many breeders attach more importance to that fact than in reality it merits, however it may naturally add to the *prestige* of any sire. 'Longbow,' to my mind, was *about the best* mile-horse under heavy weight of modern days; and as for length, power, and muscle, combined in one animal, I never saw his equal, without that top-heavy appearance so common in large horses.

### 'Drumour.'

A chestnut horse, ten years old, by 'Weatherbit,' or 'Big Jerry;' dam 'Elspeth,' by 'Irish Birdcatcher,' 'Blue Bonnet,' by 'Touchstone.' This horse in different races proved himself a very good one indeed, and ought to be successful at the stud; yet it seems almost impossible that the number of sires at present at the service of the public can all receive support, taking into consideration the comparatively small number of mares, coupled with the fact that the picked sires obtain such large patronage, consequently very many good animals, such as 'Drumour,' are frequently passed over; still, I fancy breeders might do worse than give him a chance. His grandam was 'Blue Bonnet,' winner of the Doncaster St. Leger.

### 'Dundee.'

A bay horse, six years old, by 'Lord of the Isles;' dam 'Marmalade,' by 'Sweetmeat;' grandam 'Theano,' by 'Waverley,' out of 'Cherub,' by 'Hambletonian.'

It almost amounts to absurdity to attempt to eulogise the merits of this animal; or, as Shakespeare says,—

"To gild refined gold, to paint the lily;"

whether as to his breeding, shapes, or quality as a race-horse, for each and all must be fresh in the memory of the sporting public; still, a slight "refresher" may not be out of place. His blood speaks for itself, and has told its tale; which, with such an example, must dictate to the breeder the prudence of following in the same course. His fine size, racing symmetry, with substance, without a particle of coarseness, plenty of length, temper, and soundness, cannot be questioned; his performances at two years old, over the shortest as well as the longest courses contested by two-year-olds; contending against horses of all ages and of proved speed; such as 'Ment-more,' 'Maggiore,' and others; racing under penalties, and winning with ridiculous ease, as well as proving his gameness, by doing so after a dead heat with a three-years-old, and so far beyond the usual T. Y. C. distance, must stamp him as one of the best and stoutest horses of modern days. It is remarkable that he always had to contend with first-class animals, the very cream of the year, and over their own courses; for, whether it was 'The flying Little Lady' over her short half-mile, or the old ones over their favourite distances, they all had to suc-cumb. Each were alike to him; yet the truth of the running, as well as the superiority of the class, was proved in every respect: that beautiful mare, 'Brown Duchess,' who ran second to him at Liverpool (merely getting her allowance for her sex), winning the Oaks, and otherwise proving herself, what she looks all over, about as perfect a model of perfection as, in racing *parlance*, " ever looked through a bridle." ' Dundee,' however, through meeting with an accident, which frequently attends a Derby preparation (and which the nature of the course, as well as the usual state of the ground at that period of the year, so materially adds to), only managed to get second for the Derby, which he did upon two legs; having lost the use of one at Tattenham Corner, the other at the distance, and being only just defeated, for a race run in the shortest time on record, namely, in two minutes, forty-three seconds: beating ' Diophantus,' winner of the 2000 guineas; ' Aurelian,' ' Imaus,' ' Dictator,' ' Klarikoff,' ' Atherstone,' and many others. This horse bears a great resemblance to ' West Australian,' in his general appearance and outline.

His performances were as follows:—At the Epsom Summer Meeting, 1860, he won the Woodcote stakes by three lengths; beating ' Blisworth,' ' Walloon,' and thirteen others. At Stockbridge, June same year, won a piece of plate, beating ' Damascus,' ' Mentmore,' ' Birmingham,' ' Marionette,' and five others, seven furlongs, carrying 8 lbs. extra, all ages. At Liverpool, July 1860, won the two-years-old plate, beating ' Brown Duchess,' ' Damascus,' ' Pardalote,' and ' Longshot.' At Goodwood, July same year, won the Findon stakes, beating ' Nemesis,' ' Brown Duchess,' ' Knight of St. Patrick,' and six others. At York, same year, he defeated ' Maggiore,' three years old, one mile, at one stone, and three others, after a dead heat with the former for the Eglinton stakes. At the Newmarket first October Meeting, same year, won the Hopeful stakes, defeating ' The Little Lady,' ' Walloon,' ' Queen of the Vale,' and ' Evenhand;'

at three years old, ran second to ' Kettledrum' for the Derby; after which he did not start, and was put to stud.

If there be a virtue in properly crossing particular strains of blood, the breeder of ' Dundee' evidently proved it; and if the object of others be to arrive at the summit of superiority, it would be difficult to select a better course than that of adhering to ' Lord of the Isles' and ' Sweetmeat' mares; or, taking a wider range, to ' Touchstone's' and ' Gladiator's' descendants : for, admitting there are many examples of other successful alliances (which have been already referred to), still it would be very difficult to find a better than the one in question.


## ' ELLINGTON.'

A brown horse, eleven years old, by ' The Flying Dutchman;' dam ' Ellerdale,' by ' Lanercost;' grandam by ' Tomboy;' ' Tesane' by ' Whisker,' out of ' Lady of the Tees,' by ' Octavian.'

' Ellington' proved himself a racehorse; having at two years old, when amiss, won the Champagne stakes at Doncaster, beating a good field. He also won the Derby easily, beating the unfortunate ' Yellow Jack,' ' Cannobie,' ' Fazzoletto,' and many others. In his fore-action he walked and galloped with the ' Tomboy' peculiarity. His blood is undeniable, and his family could all run. His dam being by ' Lanercost,' should add very much to his merit, as likely to get not only speedy, but stout stock. He is of a magnificent colour, the richest dark brown ; and is probably about the soundest and cleanest-legged horse at stud: and as he stands in his stall, is a perfect picture, his arms and thighs being a mountain of muscle.

'Gildermire,' his own sister, ran second to 'Governess' for the Oaks, after a dead heat; which race was won in the following year by his half-sister, 'Summerside.' His family are of a most running strain, his dam having, in fact, proved successful with every cross; and being one of the very best brood mares of modern days—so very different are cases of chance produce; one good, the rest useless.

'Ellington' gained the prize of 100 sovereigns at the Royal Agricultural Show at Battersea, in 1862; a further proof of his perfect soundness, and superiority of shape and action. His temper is most docile.

## 'FAZZOLETTO.'

A bay horse, eleven years old, by 'Orlando;' dam 'Canezou,' by 'Melbourne;' grandam 'Madame Pélerine,' by 'Velocipede.'

He is a horse of immense frame, large bone, and plenty of substance, and being descended from such fine running strains on both sides, of sire and dam, can hardly fail to get good marketable stock. His dam was one of the best mares ever foaled, and about the finest specimen of a thoroughbred mare that ever galloped : she was, indeed, a noble animal. Her performances, under enormous weights, and giving away "lumps" to racehorses—in the Newmarket handicap, for instance—besides her good second in the Doncaster St. Leger, won by 'Surplice,' after a severe contest, in which she lost a plate at the Red House, stamp her as an animal of extraordinary merit. My opinion is, that 'Surplice' never recovered the effects of that race. I perfectly remember him after it was over, and seldom saw a horse more distressed.

'Fazzoletto' himself was a very fair racehorse, for although he did not win the Derby, he ran very respectably, and I believe quite as well as was expected, for the course was by no means suited to a horse of his frame and action. He is of the same mould as 'Toxophilite,' and others of the Knowsley stamp — a fine slashing horse, but bearing a rather top-heavy, unwieldy appearance. He ought, however, to get some fine stock, which, up to the present time, he has not appeared to have done; probably from want of the chance : nevertheless, I fully expect to find some day a few slashing samples of this sire. 'Ackworth,' a fair horse, is by him.

### 'Gemma di Vergy.'

A brown horse, ten years old, by 'Sir Hercules;' dam 'Snowdrop,' by 'Heron;' her dam 'Fairy,' by 'Filho da Puta,' out of 'Britannia,' by 'Orville.'

'Gemma di Vergy' proved himself a very good horse, and his blood (being now one of the three only remaining stallions by 'Sir Hercules'), together with his well-proportioned outline, elastic springy action, and rich dark-brown colour, must render him deserving the patronage of breeders. He has, no doubt, had a fair chance, considering the number of rivals in the market, and he may yet prove more successful than he has done; for it must be borne in mind that, like other sires that have run severely-contested races, and been "wound up" to the last pitch of condition, he has barely had time to distinguish himself. He is, however, well-bred, and good-looking enough, and possibly may yet fill the place vacated by some of his ancestors. His Chester-cup running, when three years old, carrying 6 st. 11 lbs., was first-class; for although he did not win, or perhaps would not have won, having met with a *contretemps* at the distance, he would have been very handy.   He is a remarkably handsome horse, his peculiarity consisting in the shape of his hind-quarters, which he inherits from his sire; as many 'Irish Birdcatchers,' and others of the same blood, are remarkable for being rather drooping towards the tail—by no means objectionable in a racehorse, as they are generally better turned under and possess more propelling power.

## 'Gunboat' and 'Lifeboat.'

Both dark-brown horses, own brothers, aged respectively ten and nine years, by 'Sir Hercules;' dam 'Yard-Arm,' by 'Sheet Anchor,' out of 'Fanny Kemble' by 'Paulowitz,' 'Loyalty' by 'Rubens,' 'Pennyroyal' by 'Coriander.'

These two sons of 'Sir Hercules' are both worthy scions of their renowned family, having proved during their racing career very sound, good, true, and game animals, over all distances, and under heavy weights, which they are both peculiarly adapted to carry.   They are of fine size, plenty of substance all over, and their blood on the side of their dam, as well as of their sire, must strongly recommend them to breeders, more especially as their stock ought to prove valuable for any purpose, if unsuccessful as racehorses.   Few sires of the present day possess more recommendable qualities, in every shape and respect, and with fair chances they ought to distinguish themselves at stud, especially as the sons of 'Sir Hercules' ('Irish Birdcatcher' and 'Faugh-a-Ballagh ') have proved so successful as sires. Both these horses are really very fine, sound samples of the powerful thoroughbred, and bear a striking contrast to the miserable specimens with which the country is overrun.

### 'IVAN.'

A brown horse, thirteen years old, by 'Van Tromp;' dam 'Siberia,' by 'Brutandorf,' by 'Blacklock.'

The 'Lancercost' blood, so remarkable for its staying qualities, is, no doubt, well represented in the subject of these remarks; for 'Ivan' was not only himself a racehorse, which the racing records bear testimony of, but he appears likely to prove a first-class sire, having, with a very limited chance, produced 'Union Jack,' probably about the best of his year. 'Van Tromp' (his sire), winner of the St. Leger, Doncaster, and Goodwood cups, was a very superior animal, and it seems fortunate that the blood, which is at present so scarce, should at least have one representative so likely to uphold its character.

### 'KETTLEDRUM.'

A chestnut horse, with white marks, six years old, by 'Rataplan;' dam 'Hybla,' by the 'Provost;' her dam 'Otisina,' by 'Liverpool,' 'Otis,' by 'Bustard.'

Here is a fine sample of what may be expected from the loins of the renowned 'Old Ratty,' the hero of so many contests, resembling his sire in many respects, not alone in shape and colour, but in action and temper. He won the fastest Derby on record when he defeated 'Dundee,' 'Diophantus,' 'Dictator,' the unfortunate 'Klarikoff,' who, according to the idea of his friends, always ought to have won, &c. Any admirer of 'Stockwell' and 'Rataplan' can hardly hesitate to acknowledge the claims of this horse to every support, as from his running blood, good looks, and likeness to those animals, nothing but the great uncertainty attending such speculations can possibly interfere with his success. On both sides of sire and dam he comes from first-class strains, his dam having also produced 'Mincemeat,' winner of the Oaks, 1854. In addition to his victory in the Derby, when two years old he beat 'Dictator,' 'Phemy,' colt, 'Matador,' &c. At three years old he ran second to 'Diophantus' for the 2000 guineas, and second to that wonderful mare, 'Caller-Ou,' for the St. Leger at Doncaster—beaten by a head; and ran a dead heat with 'Brown Duchess' (winner of the Oaks), for the Doncaster cup: for which he, according to compromise, afterwards walked over. A curious coincidence, the Derby and Oaks winners in same year running a dead heat. This horse having previously, during the same week, ran so severe a race for the St. Leger, and carrying a penalty, is further evidence of his stoutness.

## 'KING TOM.'

A bright bay horse, with white marks, thirteen years old, by 'Harkaway;' dam 'Pocahontas,' by 'Glencoe,' out of 'Marpessa,' by 'Muley.'

As to his colour, it is said he is a bay, but according to my recollection he is a chestnut. In comparing his breeding with that of his half-brothers, 'Stockwell' and 'Rataplan,' it may not be out of place to mention an oversight made by some parties with regard to their

breeding, viz. that picking one in preference to the
other amounts almost to making a distinction without a
difference, which a reference to their pedigree will show.
In the first place, ' Stockwell' and 'Rataplan' are by
' The Baron,' by 'Irish Birdcatcher;' dam ' Echidna,' by
' Economist.'   ' King Tom' is by ' Harkaway,' by ' Economist;' therefore the latter is, in point of fact, almost
full brother in blood to the " two brothers."

    Then, suppose we take 'The Baron' and ' Harkaway'
on their merits, not only as racehorses but in every
respect as animals, and without prejudice view them
and their performances even as sires,—how can any
judge, who really has a recollection of them, conscien-
tiously pronounce 'The Baron' was a superior animal
to ' Harkaway ?' for, although the latter did not win the
St. Leger (for the best reason in the world—because
he was not in it), which invariably adds so much to the
*prestige* of any animal, in my humble opinion he was
not only a far better horse, but "proved" himself superior, as well as quite as good a sire.   It is all very
well to remember 'The Baron's' victories, and in doing
so to forget those of the animal that proved himself
about the best horse in the memory of the present
generation.    Leave ' Stockwell,' ' Rataplan,' and their
descendants out of the question, and then let the admirers
of ' The Baron' furnish a list of his descendants to boast
about, notwithstanding the extraordinary chances he had
for many years, and contrast them with the despised and
forgotten old ' Harkaway,' who reigned as the " King of
the Curragh" in his day; vanquishing every opponent
that came in his way, distancing racehorses of first class
in England as well as in Ireland, where there really were

first-class horses, and yet died, comparatively speaking, without a chance : yet left 'King Tom' as a reminiscence of his quality, as well as 'The Irish Queen' (dam of 'Sweetsauce,' probably the best horse of his year for any distance), 'The Horn of Chase,' 'Chaseaway,' 'Blucher,' 'Peep-o'-Day Boy,' 'Idleboy,' 'Ballinafad,' 'Sabroan,' &c.

Many of 'Harkaway's' daughters have proved most successful at stud. Amongst others, 'Thorn,' dam of 'Sprig of Shillelagh,' &c., 'Queen Bee,' &c. It has been hinted that there was a "flaw" somewhere in 'Harkaway's' pedigree, on his dam's side. All I can say on that point is, I should like to be owner of a few like him, bred with a similar flaw, taking into consideration that this animal could run any distance, carry any weight, and distance racehorses. It may not be generally known, that such was the *furore* caused by his extraordinary performances, that he was exhibited at a very stiff figure in Dublin for a considerable time, and vast numbers went to see him, believing him to be one of the "wonders of the world."

I remember during his racing career, amongst his performances, he was only just defeated for a four-mile Queen's plate by a horse of merit, called 'Bonté Bock,' who got about a quarter of a mile start, 'Harkaway' having been late at the post, according to the rules; his eccentric owner, however, calling out to his jockey, "Go after him; you can give him to the Hare Park and win;" a distance of about one mile and a half. The issue was well contested.

As to 'King Tom's' individual merits, he is no doubt a hopeful son of a worthy sire; he is as good-looking a horse as any *connoisseur* can wish to look at ; the thickness and substance of his frame leading some to fancy that he is a shorter horse than he really is, for when in racing form he hardly presented that appearance.

'King Tom' has proved more fortunate and successful as a sire than as a racehorse; which is proved by his 'Old Calabar,' 'Wingrave,' 'Queen of the Vale,' 'King of Diamonds,' 'Mainstone,' 'Breeze,' 'Prince Plausible,' 'Tomato,' 'Linda,' &c. If his stock have up to the present time contended for short races, that fact may be attributed to the simple reason that about one half the owners and trainers sometimes underrate what their horses really can do, and how long they can stay; and in many instances, because when half fit, or when two years old, they happen to show speed, confine their engagements to short courses, whereas in many cases their real *forte* is staying: yet they are very frequently condemned as half-mile split-tails, when perhaps four miles over the Beacon Course would suit them better.

With regard to the crosses which would best suit 'King Tom,' one patent fact must convince the reader that 'Sweetmeat,' 'Irish Birdcatcher,' or 'Touchstone,' mares (although the latter, on the "in-and-in" principle), would be most likely to hit successfully; and the daughters of this splendid animal, if properly bred on their dam's side, ought to, and no doubt will prove, most valuable as brood mares: especially bearing in mind that they will inherit the blood of that mine of value, the queen of brood mares, 'Pocahontas;' and if crossed with the sons of 'Sweetmeat,' should turn out most profitable to their owners. As instances in favour of the "mixtures" referred to, we have 'Sweetsauce,' an extraordinarily good horse for any distance, and 'Dundee,' a second edition of the former, besides numerous others; but if excellence be the object of the breeder, one would fancy the above samples, with 'Sweetmeat,' &c., ought to satisfy any "epicure." If a deviation were made, it might be desirable to try the 'Sir Hercules' or 'Irish Birdcatcher' sires with 'King Tom' mares; such as 'Saunterer,' 'Wonnersley,' 'Lifeboat,' or 'Gunboat,' in favour of which we have 'Stockwell' and 'Rataplan.' 'Citron' was another instance in favour of the 'Sweetmeat' and 'Economist' alliances, for she was an animal for any distance, seldom equalled if (in my opinion) ever excelled : a fact which is known to very few, for the simple reason that, after a few brilliant performances, she met with an accident.

### 'KNIGHT OF KARS.'

A bay horse, ten years old, by 'Nutwith,' dam 'Pocahontas' (also dam of 'Stockwell,' 'Rataplan,' 'King Tom,' &c.), by 'Glencoe;' grandam 'Marpessa,' by 'Muley.' The fact of this horse being one of the very superior sons of the above mare should alone recommend him to the notice of breeders, and render him a most likely stallion to prove serviceable at stud. But he has likewise all the other qualifications, as far as size, substance, shape, power, and colour, &c. He was a racehorse; and although never "up to the mark" in condition, his performances were of the first class, and there is no reason why his career at stud should prove an exception to the remarkable success of his half-brothers.

Amongst those horses who had to succumb to him were 'Gamester' (winner of the St. Leger), and 'Ignoramus;' and his race with 'Saunterer' at Doncaster, which was a very well-contested and near one, indeed stamps him as a horse of very high merit, and, as a sire, most likely to get valuable stock for all purposes. 'Nutwith' (his sire) won the Doncaster St. Leger, beating, amongst others, the renowned 'Cotherstone,' winner of the Derby, 2000 guineas, &c., and more money in stakes than any three-years-old on record. 'Nutwith' did not start for the Derby.

### 'LEAMINGTON.'

A dark-brown horse with a white star, eleven years old, by 'Faugh-a-Ballagh;' dam by 'Pantaloon;' her dam 'Daphne,' by 'Laurel.'

He is one of the best representatives of what a race-horse ought to be, with great length, racing points all over, and wonderful propelling power, the shape, muscular power, and position of his hind-quarters being perfection, and such as cannot fail, at a glance, to strike the eye. In his general formation and appearance he somewhat resembles his sire, and in many respects 'Buccaneer,' although more commanding in his general style; being, in fact, a perfect specimen of the fine slashing racehorse — just what might be expected from his relationship to 'Pantaloon,' whose descendants invariably present such an appearance. His sire's fame is world-wide; indeed, by many he is believed to have been the best horse ever foaled — a question which no doubt admits of serious consideration, as well as doubt. Be that as it may, that 'Leamington' looks all over a fine model of a racehorse; that he proved himself 'which is better than mere appearance, which is often a fallacy) a genuine first-class animal, as well as a perfectly

sound and wear-and-tear one, cannot be denied, for although his name is not recorded amongst the Blue Riband
or St. Leger winners, his performances justly entitle him
to the confidence and support of breeders, which would be
badly rewarded by patronising, as a rule, some of the
winners of those great events; for although the Derby,
Oaks, and St. Leger, may be generally very good tests of
quality, it by no means follows that there are not, in
many instances, far better horses of the year than the
winners — some never even entered.   Taking into consideration the fact, that the country has lost his sire,
'Leamington' must be looked to as the most promising
son of that renowned animal best qualified to fill his place,
and uphold untarnished the *prestige* of his ancestors;
which, with a fair and reasonable chance, there can hardly
be a doubt he will do, for he is from head to tail a "noble,
fine animal," and one most likely yet to stand at the same
figure as his relative, 'Irish Birdcatcher.'   The fact that
he has the 'Pantaloon' blood in his veins is an additional
recommendation, for it is questionable if there flows in any
animal better : the very best runners, and the grandest
specimens of the noble thoroughbred, are descended
from 'Pantaloon.'   I believe that 'Leamington,' as
viewed upon a racecourse, walking with his majestic
yet steady air, presents at once the appearance of the
most level-made, lengthy sample of a racehorse, that
we have seen for many years; every shape and point
being where they should be, and his "propellers" always
doing their duty; being placed so beautifully for the
purpose.   Although his career at the stud has, as yet,
but commenced, he has begun well; and is, to my mind,
certain to finish better.   The running of 'Fille de l'Air,'

Q

that extraordinary mare (a daughter of 'Faugh-a-Ballagh'), is further proof in favour of the chance of 'Leamington' proving successful.

## 'LORD OF THE ISLES.'

A bright bay horse, with the usual white marks of his family; eleven years old, by 'Touchstone;' dam 'Fair Ellen,' by 'Pantaloon;' out of 'Rebecca,' by 'Lottery,' 'Cervantes,' 'Anticipation,' by 'Beningbrough.'

'Lord of the Isles' blood cannot be excelled on either side, being a combination of speed and stoutness. In his general *contour* he bears a striking contrast to his successful opponent, 'Wild Dayrell.' Probably no two animals of such merit could be more dissimilar, not only in shape and general appearance, but in action; proving the truth of the opinions so frequently expressed, that "they run in all shapes." He is neither a lengthy nor a short horse, but of average size; his 'Touchstone,' muscular quarters, his well-knit frame, being well-proportioned, give him the appearance of a fair-sized, compact animal: he possesses the usual propelling power of the 'Touchstones,' and in other respects the quality requisite in the racehorse. During his career he won the 2000 guineas, beating 'St. Hubert,' who had been highly tried, and others, after one of the closest and most severely-contested races on record; then meeting for the Derby that splendid specimen of the racehorse, the home-trained 'Wild Dayrell;' besides a few animals of very moderate pretensions. It can hardly be questioned that, on that day, 'Lord of the Isles' was not up to the mark, but rather beyond it; for I perfectly remember that, early on that morning on the Downs, when he was at exercise, and endeavouring to come down the hill at Tattenham Corner, he reminded one more of the action of a rabbit than of a racehorse; regularly "stumped up," evidently suffering from sore shins, as well as from not having recovered the effects of his severe race for the 2000 guineas. He could not move on that day, nor did he appear quite at home during the race; being beaten for second place by an animal not far removed from a "leather-plater." However, whatever may have been his qualities as a racehorse, he has given good proof of his value as a sire; for, before we can give the preference to others, we must see the superior of 'Dundee.' Besides which he has furnished 'Scottish Chief,' 'Donna del Lago,' and others; in addition to which there are numbers of his young stock of great promise, fine size, and racing symmetry; and it will surprise me if we do not hear of his ranking even higher than at present in the estimation of breeders. 'Dundee' is a proof in favour of crossing with 'Sweetmeat' mares.

### 'THE MARQUIS.'

A bay horse, six years old, by 'Stockwell;' dam 'Cinizelli,' by 'Touchstone;' grandam 'Brocade,' by 'Pantaloon,' out of 'Bombasine,' by 'Thunderbolt.' A greater proof of the success of the "double mixture" can hardly be found than in 'The Marquis;' showing the excellence of the 'Touchstone' and 'Birdcatcher' (or rather 'Stockwell') cross; which is also borne out so plainly in the cases of 'Asteroid,' 'Audrey,' and others.

As to size, shape, and power, he is good-looking enough

for anything ; his performances were first-class ; and there can be little doubt he will distinguish himself as a sire.

During his racing career, at two years old, he won the Champagne stakes at Doncaster, beating 'Feu-de-Joie' (winner of the Oaks, 1862), 'Impératrice' (second for the Oaks, 1862), and others. At three years old he won the 2000-guineas stakes, defeating 'Caterer,' 'Alvediston,' 'Wingrave,' &c. ; and the Doncaster St. Leger, beating 'Buckstone,' 'Hurricane,' 'Johnny Armstrong,' 'Carisbrook,' &c. He was beaten only once, and then by a neck for the Derby, by 'Caractacus.'

## 'MUSJID.'

A dark-brown horse, nine years old, by 'Newminster ;' dam 'Peggy,' by 'Muley Moloch ;' grandam 'Fanny,' by 'Jerry,' 'Fair Charlotte,' by 'Catton,' 'Henriette,' by 'Sir Solomon.'

As a winner of the Derby, this horse is entitled to the highest consideration ; for he not only won that race like a racehorse, but by sheer gameness. He is a very fine animal indeed, and his pedigree comprises some of the real old, although neglected, strains ; and differs very much from the everlasting, every-day, hackneyed names we read. It is refreshing to see a Derby won now-a-days by an animal with a pedigree like that of 'Musjid.' He has in his veins the blood of those two wonderful mares, 'Beeswing' and 'Alice Hawthorne ;' no mean recommendations : and it is highly probable he may prove as successful at the stud as he did on the racecourse.

## 'NEWMINSTER.'

A bay horse, seventeen years old, by 'Touchstone;' dam 'Beeswing,' by 'Doctor Syntax;' grandam by 'Androssan.'

The merits of this horse are so thoroughly known and tested, both on the racecourse and at stud, that any attempt to expatiate upon them would be like sending coals to Newcastle. He is a gem of the first water; and his fine son, 'Lord Clifden,' the *beau idéal* of a racehorse, alone entitles him to the high estimation in which he is held, as well as the fact that he is descended from the renowned 'Beeswing.' He won the Doncaster St. Leger in the commonest of canters; and his success at stud is so universally known, that it is merely necessary to leave the public to choose between him and his rivals in the "known world,"— 'Stockwell,' 'Rataplan,' 'King Tom,' 'Voltigeur,' 'Lord of the Isles,' and 'Wild Dayrell.'

## 'NUTBOURNE.'

A dark chestnut horse, eight years old, by 'The Nabob;' dam 'Princess,' by 'The Merry Monarch;' grandam 'Queen Charlotte,' by 'Elis;' great grandam by 'Tramp,' out of 'Fillagree,' by 'Soothsayer,' out of 'Web,' by 'Waxy.'

Who that ever saw a splendid specimen of the horse can deny that 'Nutbourne' is one? Take him "all in all" as he stands, and where is his superior, in a general point of view? A mountain of muscle—a racehorse even at two years old—an age at which such large animals seldom show their real form, he met and beat, not only the best horses of his year, but probably the best that had appeared for years. His blood is a mixture of most of the best crosses for speed and stoutness. His performances at two years old were sufficient to stamp him as a first-class animal; and judging of them in a fair and impartial manner, and taking into account the fact, that such large horses generally improve wonderfully from two to three years old, as also that when running for the Derby he met with an accident when "pulling double" (the race being won by 'Thormanby,' whom he had previously defeated at two years old), leaves little doubt that, but for the *contretemps* referred to, 'Nutbourne' would have proved the most dangerous opponent to the winner. It appears strange that the medal should have been lately awarded to another, in opposition to this horse, at the show: however, "doctors differ;" and every man has an equal right to his opinion, which must be received and taken for what it is worth.

In my humble opinion, if all the horses in England were brought to an exhibition, and the medal was offered for the "finest sample of a thoroughbred stallion, for general purposes, taking breeding, *performances,* soundness, size, shape, and, in fact, all qualifications into consideration," and if a dozen experienced judges in such matters were appointed, 'Nutbourne' would wear the medal in opposition to any other, and be justly entitled to do so. I have previously referred to this horse's blood, under the head of 'The Nob.'

### 'OLD CALABAR.'

A bay horse, six years old, by 'King Tom;' dam by

'Picaroon;' grandam 'Jemima,' by 'Count Pozzo;' 'Mrs. Suggs,' by 'Crispin.'

The best son of 'King Tom,' with plenty of size and substance, *looking* all over what he *proved* himself, "a first-class racehorse;" but, having met with an accident, one whose career was but short on the turf. He may some day, or perhaps now, rank amongst the *unfashionable* division, because he did not *win the Derby*, and because his owner may not deem it worth while to take the trouble of convincing breeders that they would be studying their own interests in giving him a trial. To my mind he is one of the most promising untried sires of the present day, grounding my opinion on the merits of the animal in every respect, as to breeding, size, shape, and performances; and believing that (bar accidents) he would have proved himself the best of his year at three years old, which he did as a two-years-old. How many others untried are more likely to prove stars at stud, notwithstanding the medley of *unfashionable* names which his pedigree displays?

At two years old 'Old Calabar' was not beaten, having won four times; namely, the Triennial produce stakes at Newmarket; first October meeting, beating 'Hurricane' and others; the Clearwell, beating 'Wingrave' (also by 'King Tom'), 'Knight of St. Michael,' and others; 'The Criterion,' beating 'Nottingham,' 'Alvediston,' 'Wingrave,' 'Feu-de-Joie,' 'Zetland,' 'Bertha,' and others; and the Glasgow stakes.

## 'ORLANDO.'

A bay horse, twenty-four years old, by 'Touchstone,'

out of 'Vulture,' by 'Langar,' out of 'Kite,' by 'Bustard.'

The sire of many speedy animals, and very few stout
horses, 'Teddington' and 'Impéricuse' being the best,
having had for a number of years the best chance of
any stallion that ever lived, he won the Derby in the
memorable year of 'Running Rein.'


### 'OULSTON.'

A bay horse, thirteen years old, by 'Melbourne;'
dam 'Alice Hawthorne,' by 'Muley Moloch,' out of
'Rebecca,' by 'Lottery.'

'Oulston' is a peculiarly-shaped animal, a sensible,
" steely-looking gentleman," without a particle of lumber,
and with particularly black points, resembling " the old
mare" in many respects, as well as colour; although
very round-ribbed, with a deep girth, like his dam,
rather shallow, which, however, is made amends for
by a good back and loins; his hind-quarters well placed,
which, like 'Old Alice,' he could use to perfection :
he has very lopped ears, for which many of the 'Melbournes' are remarkable,—'Sir Tatton Sykes,' &c., to
wit.

As a racehorse he was of the highest order, his
performances being of so brilliant a character that he was
sold for (about the highest price on record) 6000 guineas,
which, no doubt, does not at all times prove the intrinsic
value, so much frequently depending upon the ideas,
sagacity, and talents of buyers and sellers. I have sold as
good for a sixth of the price, with wonderful *expectations*
of contingencies. However, it cannot be denied that

'Oulston' was a first-class racehorse, and might have probably shown more of his quality had he not been occasionally seized with a slight distemper on the eve of his engagement.   His running was very much of the in-and-out kind, still many circumstances occur in racing which can account for this; for instance, amongst others, horses are not and cannot be always fit, and up to the mark.   As far as 'Oulston' is concerned it is very questionable if he was not the best horse of his year, not even excepting the great 'Wild Dayrell;' and it was very fortunate that the former was not up to the mark, and that 'Rifleman' was not "primed and loaded," on the Derby day, or he might have taken down 'Wild Dayrell's' colours.   'Fandango' and 'De Clare,' being *hors de combat*, also made the coast pretty clear for the great gun.   It is extraordinary that this horse has not been better supported at stud; but, as before stated, it is impossible to give all a chance, where there are so few mares in proportion to sires.   And probably a report as to an infirmity in 'Oulston' having got wind, has more or less injured him; although it has been, and is stated by proper authority, that there are no grounds for such a rumour.   There is one patent fact, viz. that his sire and dam, and their descendants, in fact all his family, are remarkable for their soundness in that respect, but are likewise so for their great staying powers; and 'Oulston' himself could run any distance, and distance horses: and even, if there were any ground for such a report, the infirmity, if it does or ever did exist, must have had its origin in some distemper, and, therefore, cannot be here-ditary or constitutional.   This horse has literally had no chance at stud, and it will by no means surprise me to see his stock some day astonishing the talent, and bringing up reminiscences of bye-gone days.

### 'Petruchio.'

A chestnut horse, eight years old, by 'Orlando;' dam 'Virago,' by 'Pyrrhus the First;' grandam 'Virginia,' by 'Rowton;' great-grandam 'Pucelle,' by 'Muley.'

Here is a son of the great 'Orlando,' as well as the selling stake 'Virago,' whose qualities the talented Mr. Topham appeared to estimate beyond leather-plating form, when he treated her to seven stone, three-years-old, for Chester cup; yet how truly was his well-known judgment displayed!—the mare being considered by her trainer about ten pounds better than 'Crucifix.' It appears to me that 'Petruchio,' who was a very good-looking young one, with capital propelling power, ought, with a fair chance, to prove more successful at stud than he did otherwise; for if high pedigree can recommend any horse, surely he possesses it. His success at stud (if he has grown as one would have expected, and is sound, &c.) would not be a greater cause for wonder than his failure as a racehorse; for, like many others that were not racehorses, yet turned out valuable and successful at stud, he may yet prove " that blood will tell."

### 'Prime Minister.'

A brown horse, seventeen years old, by 'Melbourne;' dam 'Pantalonade,' by 'Pantaloon;' grandam 'Festival,' by 'Camel.'

This horse possesses many qualifications to recommend him to the notice of breeders; amongst others, his being a combination of the most running and winning strains of blood, inferior to none. As a racehorse he was very highly tried previous to the Derby, and was backed by his partisans to win as much money as would have purchased half Manchester. He is a very level, racing-like animal, and has proved as successful at stud as could have been expected, taking into consideration that he has not had as much patronage as others less worthy of it, although more "cracked up." He has, however, proved his high quality as sire of several good animals; amongst others, 'Farfalla,' 'Lord Burleigh,' 'Light,' 'Lustre,' 'Pastime,' 'Sporting Life,' 'Tesane,' &c., and doubtless will yet add to the number.

The fact of possessing the 'Pantaloon,' 'Melbourne,' and 'Camel' mixtures, should especially recommend him to the notice of breeders. His stock appear remarkably sound, wiry, and racing-like in every respect.

### 'RATAPLAN.'

A chestnut horse, with white marks, fifteen years old (own brother to 'Stockwell'), by 'The Baron;' dam 'Pocahontas,' by 'Glencoe, out of 'Marpessa,' by 'Muley.'

His iron constitution, with the strength of an elephant and temper of a lamb, together with the fact of his having proved himself a racehorse under any weights, and for any distance, must render 'Rataplan,' commonly called 'Old Ratty,' invaluable for stud purposes; and having begun well with his 'Kettledrum,' 'Miner,' 'Tattoo,' &c., there can be little doubt he will play to

perfection with other instruments a more prominent part
ere long. What a temper! It sometimes required the
aid of the late Mr. Hibburd's cob (so well known to
frequenters of the racecourse) to set him going; but
once the steam was on, it was like "Hell-fire Jack's"
engine from Didcot to London, whose fondness for pace
and keeping time is so well known to travellers on the
Great Western. The extraordinary number of races (and
those over the most severe courses) won by this wonderful
horse, and his successful career, is in a great measure to
be attributed to the almost incomparable, and certainly
unexcelled experience, skill, and talents of the "Squire
of Wantage;" by whose instructions, and under whose
practised eye, his trainer no doubt brought 'Old Ratty'
out much oftener than he could have appeared, if under
the usual style of continual racing and training at the
same time : for this horse, like the everlasting old 'Fisherman,' together with others, after the Wantage style, was
trained by running races and winning money, or, as the
adage is, "killing two birds with one stone," getting,
however, that rest at intervals which some wretched worn-
out skeletons never do, whose trainers think it necessary,
whether at home or abroad, to keep them continually
galloping: the consequence being, their early retirement
to the stud or Hansom. 'Rataplan' is a perfect rock of
strength, a little plain in the shoulders, rather short in
his fore-action, which stayers frequently are (flash-goers
the contrary). He is not so tall or commanding in his
appearance as his brother 'Stockwell,' nor did he probably possess his speed : still, it is hardly possible to exaggerate his good qualities; for however frequently we
find holiday-flyers, or even stayers, still we have seldom,

if ever, seen a more genuine sample or combination of all the requisites in the racehorse than he affords. " He was all, and always there." It is but reasonable to suppose, that having already produced a Derby winner, besides others that could run, and not only stay, but show great speed, his reputation as a sire ought to, and doubtless will improve; especially taking into consideration the fact that he had been so long in training, and consequently had done so much work, and that time must, as with most horses when put to stud, prove beneficial to ' Rataplan.'

Amongst his performances, his winning the Manchester cup, literally in a canter, with such a heavy weight (9 st. 4 lbs. as well as I remember), beating ' Typee' and other first-class animals, stamps him as one of the best horses of modern days. He is, indeed, one of the few samples of a " genuine racehorse:" his stock must prove useful for general purposes, and I believe it is a question admitting of some doubt whether he will not yet rival his brother, especially as a sire of stayers.

### ' St. Albans.'

A chestnut horse, eight years old, by ' Stockwell;' dam ' Bribery,' by ' The Libel;' grandam ' Splitvote,' by ' St. Luke;' great-grandam ' Electress,' by ' Election.'

As to size, shape, and other recommendations, it would be difficult to select a better style of racehorse; he is one of the several fine specimens of his sire's stock, most of which appear to stay as well as show speed. This horse very much resembles his sire in his style and shapes, and is another of the many proofs of the value of my favourite blood, ' Pantaloon.'　Whether judging from his blood, shapes, or performances, there are few of the untried stallions of the present day can compete with ' St. Albans,' or are so likely to prove successful at stud.　His performances were really first-class, having won the Metropolitan, the Chester cup, and the Doncaster St. Leger.　As to the Chester cup, my conviction is that he would have won with two stone more on his back, especially if well held together, which no boy living of five stone could do round such a course as Chester, and for such a distance.　' St. Albans' resembles his sire very much, and if he fail to get racehorses it will be one of the extraordinary anomalies of breeding.　He is a fine animal, and destined, in my opinion, to be the ' Stockwell' of a future day; at least, of those of that sire's sons at present at stud, for if ' Asteroid' were a rival he would be a very dangerous one.

### ' STOCKWELL.'

A chestnut horse, sixteen years old, by ' The Baron ;' dam ' Pocahontas,' by ' Glencoe ;' grandam ' Marpessa,' by ' Muley.'

The merits of this superb sire hardly require comment; he stands at fifty guineas (and likely shortly to do so at one hundred guineas), and his subscription full every year is demonstrative of his superiority.　As instances of his quality we have the four St. Leger winners, ' St. Albans,' ' Caller-Ou,' ' The Marquis,' and ' Blair Athol ;' as also his sons and daughters, ' Audrey,' ' Asteroid,' ' Stockade,' ' Comforter,' ' Thunderbolt,' ' Bertha,' ' Caterer,' ' Bathilde,' ' Lady Augusta,' and numerous others.　The cross between this horse and ' Touchstone' mares bears extra-

ordinary proof of the excellence of the double cross of the
'Whalebone' blood, as also that with 'Pantaloon,' the
latter being, to my mind, "an improvement to any strain."
Nothing is more absurd than commenting on subjects
which are universally proved and known, therefore any remarks from my pen upon the merits of 'Stockwell' are
unnecessary, especially as they have been so well and ably
described elsewhere, by competent judges and writers upon
the sires of the day.

### 'SURPLICE.'

A bay horse, twenty years old, by 'Touchstone;' dam
'Crucifix,' by 'Priam.'

One of the most striking, as well as extraordinary instances of the lottery of breeding, is furnished in this
horse, a winner of the double event, Derby and St. Leger,
possessing all the size, shape, and power, of the true racehorse, even on a grand scale; plenty of length and substance; a pedigree composed of the very essence of good
blood; still a perfect failure as a sire. He beat in the St.
Leger that magnificent mare 'Canezou' (who lost a plate at
the Red House), piloted by that splendid horseman the late
Frank Butler — notwithstanding whose determined efforts
poor "Nat" succeeded in landing the yellow jacket, although
many were of opinion that, but for the loss of her plate,
the mare would have won. But why has this horse been
so unsuccessful at stud? One of the causes, I fancy,
must have been, that he did not get the right stamp of
mares, for his stock were generally immense, tall, leggy,
unweildy, and top-heavy in their appearance, and even
when walking seemed to drag their legs behind, as if they
did not belong to them; they were mostly like a giblet

pie, "all legs and wings," with action like an ostrich, yet without the speed of that bird. The fact is, it so happened that 'Surplice' was unfortunate in the class or stamp of mares sent to him; probably he had not nice, compact, average-sized mares, like the 'Sweetmeats' or 'Irish Birdcatchers,' for it appears to me absurd to suppose that an animal of his class, quality, and other recommendations, could possibly, after a period of nearly fifteen years, pass into oblivion without having more brilliantly upheld at stud his great fame as a racehorse; and even now, in my opinion, if some of those experienced breeders, who take care of their stock and spare no expense, would send some of the class of mares referred to, there is little doubt they would find that there is yet a chance for 'Surplice,' especially if crossed with 'Sweetmeat' mares: for 'Dundee' bears proof in favour of the 'Touchstone' and 'Sweetmeat' alliance; and the facts that the 'Sweetmeat' mares are shortish, compact animals, whilst 'Surplice' is as long as a man of war, and that the latter's blood is that of 'Lord of the Isles' (sire of 'Dundee'), are so strong that I should then, in case of failure, despair of the magnificent son of 'Touchstone' and 'Crucifix' ever raising himself from his present fallen position. The fact is, mistakes are made in breeding from such large horses with very large mares.

'Flax,' dam of 'Queen Bertha' (winner of the Oaks), is by 'Surplice,' and I fully expect to see his sons and daughters prove, like those of 'Bay Middleton' and others, more valuable at stud than on the turf. Writing about this horse calls up reminiscences of that worthy specimen of the sensible English trainer, the late respected Mr. Isaac Day, with his large cigar and his string of ostriches.

## 'THORMANBY.'

A chestnut horse, eight years old, by 'Melbourne,' or 'Windhound;' dam 'Alice Hawthorne,' by 'Muley Moloch;' grandam 'Rebecca,' by 'Lottery,' 'Cervantes,' 'Anticipation,' by 'Beningbrough.'

It has been said that there is an eloquence in silence, which course might be adopted with regard to this horse, as he has already given such proof of his quality. It is hardly necessary to remind those who have made racing their source of amusement, profit, or loss, of the great merits of 'Thormanby.' If ever there was an instance of genuineness, and a sterling proof of the value of keeping to the running strain, here is one furnished in the son of the old mare—a recollection of whose name, and wiry, racing-like form, will never die while racing exists; her deep girth, her racing shape, with length and strength where they ought to be in the racehorse, unencumbered by a particle of lumber, have been handed down by Nature to her son 'Thormanby.' She was not a float-horse, nor was she framed like one; she was formed as a weight-carrying racehorse ought to be (for action carries weight); and if she was light in her back ribs, many of the best and longest runners have been so likewise, and especially animals that have fine propelling power and hind action, which is seldom seen with those well ribbed-up, however strong they may be for other purposes.

As to the "double sire," 'Melbourne,' or 'Windhound,' there can hardly be a question on that point. Where are the "lopped Oulston ears?" How many 'Melbournes' were chestnut? What colour was 'Panta-

loon' (sire of 'Windhound')?    Are not the 'Melbournes'
invariably plain, and broad across the hips (no doubt all the
better for the latter)?    But who can find the least resemblance between 'Oulston' and 'Thormanby' in any respect whatever, except that they were both just what
might be expected from such a dam—first-class animals?
Although it by no means follows that because the mare
had been last served by a particular horse, he must necessarily be the sire of the produce, still it generally is the
case; and it must be borne in mind, that at the time the
question of the impotency of 'Melbourne' was very
much canvassed.

'Thormanby' looks all over a 'Windhound,' although
not of his colour.    Still he is that of 'Pantaloon.'    His
good length, especially from hip to hock, he takes from
his dam.    His sensible ideas of taking things as they
came were wonderful; when saddling, running, weighing,
or feeding subsequently, all appeared alike to him.    He
could run at two years old as often as he was required,
which was frequently; he could let his opponents get a
good start and beat them afterwards (he was not a very
good beginner); he ran and won over all distances, upsetting pots that had been boiling, the contents of some
having been almost eaten before they were cooked.    Many
of the wizards imagined he was "a perfect cure;" but they
should not have been surprised, for the simple reason that
the second in the Derby ('The Wizard') beat all the rest: if
he did stop, he was defeated by an animal that would make
"many a Derby winner remember the Derby day," and
refuse his corn in the afternoon.    How can this horse fail
to be a successful sire?    No matter from what animal
he is descended, whether from the sire of 'West Austra-

lian' or 'Windhound' (the son of 'Pantaloon'), his dam was 'Alice Hawthorne.'

During his racing career he had a medley of good and bad luck. He always met first-class animals, and at times was not up to the mark. Where is the trainer who can ensure and keep a horse fit? Soon ripe, soon rotten. Constant dripping will wear a stone. What splendid contests may the lovers of the turf not anticipate between the produce of the very animals which contended with this horse during his racing career—'Thunderbolt,' 'St. Albans,' 'Nutbourne,' and 'Buccaneer?' Like 'Voltigeur,' 'Thormanby' was rejected by all the *connoisseurs* during the yearling sales at Doncaster, and was subsequently purchased by his trainer, Mr. Matthew Dawson, for three hundred and fifty guineas: a proof of his well-known judgment and experience.

### 'THUNDERBOLT.'

A chestnut horse, eight years old, by 'Stockwell;' dam 'Cordelia,' by 'Red Deer;' grandam 'Emilia,' by 'Young Emilius;' great grandam 'Persian,' by 'Whisker.'

Here is a horse of extraordinary power and substance, with most wonderful loins and quarters; a perfect rock of strength, and, in fact, as fine an animal as ever was foaled. During his racing career, it often struck me that an oversight must have caused his owner to confine his engagements to short distances: the reason, as well as I remember, assigned was because of some infirmity in his feet or legs, which interfered with his training. Be that as it may, for the distances he contended he was an out-and-out good animal; and the example of 'The Baron,' in

his races for the Madrids, with 'Highwayman' and
others, often led me to fancy that the same slight cause
may have interfered with 'Thunderbolt,' viz. his feet;
and that the addition of the piece of leather, which Mr.
Watts so judiciously and successfully applied in 'The
Baron's' case, might have had the desired effect with
'Thunderbolt:' for, to judge from the latter's breeding,
shapes, and other recommendations, one would be slow to
question his powers to stay any distance.    He certainly
was a first-rater for those for which he did contend,
and as he stood while his jockey was weighing in, it
would puzzle a judge to find a grander specimen of a
thoroughbred horse; and it is questionable if his superior
be amongst the untried stallions of the present day.    His
giving 29 lbs. to 'Brown Duchess' (who in Doncaster
cup ran a dead heat with the winner of the fastest Derby
on record), for the year and sex, at Newmarket, so late as
the month of October, showed his wonderful speed, besides
his other very great performances: such as in October,
1860, when he defeated 'Buccaneer' and 'King of Diamonds,' in the Select Stakes at Newmarket, one mile; at
Warwick Spring Meeting, 1861, where he won the Trial
stakes, one mile, with 8 st. 9 lbs., beating 'Lady Clifden,'
'Lifeboat,' 'Twilight,' and others.    The Stamford plate,
Newmarket, three-quarters of a mile; beating 'Stampedo,' 'Maggiore,' 'Twilight,' and others.    His defeating
'Fravola,' 'Maggiore,' and others, easily; and his running at two years old, in which year there were many
first-class horses, such as 'Thormanby,' 'Nutbourne,'
'Buccaneer,' &c.—stamp him as a horse most appropriately
named.    There can be little doubt that, with a fair
chance, he will distinguish himself at stud.    He is, in my

humble opinion, about the best-topped horse in England (and perhaps the best we have seen, for a certain distance, for many a day) : and it would take a great deal to persuade me that his ailment was not in the feet—a second edition of ' The Baron,' "thin soles ;" which, being a heavy horse, affected him perhaps more than it otherwise might have done, especially on hard ground. His stock ought to be very fine, and with such a capital mixture of staying blood they should run as long as those of any other sire living.

### ' Toxophilite.'

A dark-bay horse, ten years old, by ' Longbow ;' dam ' Legerdemain,' by ' Pantaloon ;' grandam ' Decoy,' by ' Filho da Puta ; great-grandam ' Finesse,' by ' Peruvian.'

This horse is one of those fine slashing samples which so frequently represent the descendants of ' Pantaloon,' differing so much from other short " trussed-up ones." His sire (although a *musician*) was one of the most powerful and best horses of his day, with great length, wonderful substance ; his arms, shoulders, and thighs, a mass of muscle. His performances over the mile course, and occasionally a little beyond it, are probably unsurpassed : his winning the Stewards' cup at Goodwood, carrying 9 st. 4 lbs., beating cleverly an immense field, placing him at the top of the tree. His son ' Fox,' as he was usually called, proved himself a first-class racehorse ; for although beaten for the Derby, he ran a very good horse, although the course was by no means suited to an animal of his shape and action, he being very much of the same stamp as ' Fazzoletto,' and what might be termed " top-heavy." He is, however, a fine specimen of the racehorse, and, with such running blood in his veins, most likely to prove successful at the stud.

His dam proved her staying qualities, and 'Feu-de-Joie,' winner of the Epsom, and Yorkshire Oaks, and of the York cup, shows that although 'Longbow' himself (the most muscular, lengthy, and powerful horse I ever saw, and on short legs), was a "miler," his stock can stay; and there is no reason why 'Tox' should not furnish Derby, Oaks, or St. Leger winners, should he get a fair chance; and if ever one owner more than another deserved to win the three events, as well as breed the winners, it is the noble proprietor of 'Toxophilite,' whose success would no doubt, upon all sides, be justly hailed with ovations never before equalled, certainly never better merited, as the victory of the greatest, gamest, and most staunch sample of the noble, true, and thoroughbred sportsman. If 'Tox's' sons prove their staying qualities as well, *they will do !*

### 'Vedette.'

A rich brown horse, eleven years old, by 'Voltigeur;' dam by 'Irish Birdcatcher,' out of 'Nan Darrell,' by 'Inheritor,' out of 'Nell,' by 'Blacklock.'

The prejudices which are entertained by parties for their respective favourites are various. I candidly confess that my "weakness" is in favour of 'Vedette,' in preference to any untried sire of the present day (with one exception, 'Dundee'), if exclusively confined to racing purposes; although no doubt there are others, upon a grander and more commanding scale, and preferable, as fine specimens of the horse, in a general point of view.

My remarks are, however, principally confined to his merits as a racehorse, and my belief is, that he is good-looking enough in other respects.

With regard to his general conformation he is a very fair-sized horse, with good length, standing about 15 hands 3 inches. Whatever his other merits may be, however he may have occasionally suffered from a sort of rheumatic affection (of the nature of which I am ignorant), he made an impression on my mind which few others ever did. Whatever his "private" trials may have been (they are frequently mere moonshine), his "public" ones satisfied me as to his excellence; and I believe, without exception, the best horse I ever saw gallop two miles was 'Vedette.' I have seen old 'Harkaway,' 'Mount Eagle,' 'Skylark,' 'Irish Birdcatcher,' 'Faugh-a-Ballagh,' and many of the "stars" of bygone days, and from childhood have made horses my study. I have never been sanguine about the success of particular animals; on the contrary, have been a believer in the old adage, "there are as good fish in the sea as ever were caught;" still, when 'Vedette' was about to start I looked upon his success as a foregone conclusion. Who that ever saw true action (when set going), could exaggerate the beautiful level stride of this animal, with the "propellers" so regularly and powerfully doing their duty like a steam-engine, when his superior, yet unassuming jockey, Johnny Osborne, used to pass the winning chair, sitting as cool as a cucumber, and returning to scale amid the congratulations of the patrons of the "spots," and the public in general, with his usual imperturbable countenance, while some of his scattered opponents were straggling in, having found pursuit hopeless. True, he was not in the Derby (a lucky circumstance for the lot); even granting that, perhaps, it might not have proved exactly the course for him : but take the "bunch" that that ran in his year, and let them try conclusions with 'Vedette,' over York, Doncaster, or Newmarket racecourse, even for the Derby distance, what would he have done with the lot ? and if two miles, or "Cæsarewitch distance," he must have won, "hands down as usual." A sheet would have covered five or six of the front rank in that Derby, and what earthly chance could such horses as 'Strathnaver,' 'Anton,' &c., have had with the animal in question ? 'Faugh-a-Ballagh,' according to the usual nine-days' wonder principle, was believed by many to have been the best horse ever foaled : for my part, if it were possible that both could have met at even weights over the Cæsarewitch course in their best day, I verily believe 'Vedette' would have won, notwithstanding the wonderful opinion formed and run away with about the great 'Wonder of Erin,' a great portion of the nine-day fever originating in the fact that he was the first Irish horse that ever won the St. Leger. His success was like "Moses and Co.," the outfitters—well advertised.

'Vedette' not happening to be even entered for the Derby, must consequently, according to the opinions of some, be comparatively inferior. 'Faugh-a-Ballagh' only beat 'The Cure' by a head for the St. Leger, the latter having swerved across the course : this horse not only won his races, which were of the first order, and in which he was opposed by the best of his day, but literally walked in. Take his running in the 2000 guineas with 'Anton' and others, and that of the latter in the Derby (although a

small horse), the mile being in his favour and against
'Vedette;' then, again, 'Saunterer's' form (not in the
Derby, as he was notoriously amiss) in the Cambridgeshire,
and the Goodwood cup, ought to convince any impartial
judge that 'Vedette' was by far the best horse of his
year.   It has been asserted by some that 'Skirmisher'
was nearly as good in private—he failed to prove it in
public : for when both met over York racecourse, and
ran in different interests, the result was a very hollow
affair indeed, for my pet won in his usual style ; and
if my memory serves me, 'Saunterer' formed one in
the field.   Then if we look at 'Skirmisher's' Ascot-
cup victory, it only confirms the fact, that in the great
'Blink Bonny' year there was no horse within lengths
of 'Vedette,' especially over a reasonable distance of
ground.   If there were one near him it was 'Sprig of
Shillelagh,' an overgrown two-years-old, that when dead
amiss, coughing, and meeting a serious accident ten days
previously, during which time he was physicked and walking, beat 'Blink Bonny' at Chester.   Whatever the success of this animal at stud may be, my opinion of his
superiority as a racehorse will remain unaltered, and I
believe he will (with a fair chance), as certain as I pen
these remarks, prove more successful at stud than his
sire.   Notwithstanding the fact that his dam is by 'Irish
Birdcatcher,' my selection of blood in mares to cross
with 'Vedette' would be 'Touchstone,' 'Sweetmeat,'
'King Tom,' and 'Orlando.'


'VOLTIGEUR.'

A brown horse, eighteen years old, by 'Voltaire,' out

of 'Martha Lynn;' by 'Mulatto,' out of 'Leda,' by 'Filho da Puta.' ,

It appears strange that parties should pick out this horse to condemn and disparage, for as far as his blood is taken into consideration it cannot be excelled: his performances are a matter of record. One of the few horses that won the "double event," Derby and St. Leger, and the only animal that ever vanquished 'The Flying Dutchman.' His colour is a beautiful rich dark brown, with legs like jet, plenty of size, and racing shape; in short he has, both on the racecourse and at stud, given such unmistakable proofs of his quality that further remarks are quite superfluous. Probably those who find fault with 'Voltigeur,' or his blood, on either side, either as a racehorse or at stud, would favour the public by explaining what may be their definition or ideas of a racehorse and his blood, and what they ought to be. No doubt his stock are remarkable for their fine, improving, and staying qualities, but it by no means follows that they are so deficient in speed, although the former may be and is their admitted *forte;* and a very good one it is, and seldom found in other strains. I have previously referred to the crosses with this blood elsewhere.

### 'WARLOCK.'

A roan bay horse, twelve years old, by 'Irish Birdcatcher;' dam 'Elphine,' by 'Emilius,' out of 'Variation' (winner of the Oaks in 1830), by 'Bustard.'

The sire of this horse, and the sire of his dam, stood each at the figure of fifty guineas, demonstrating the high estimation in which they were held. He won the Doncaster St. Leger, and from his high breeding ought to prove successful at the stud. He is an average-sized animal, his great peculiarity being in his colour, which he takes from his sire ; with, however, much more of the grey or "silver" hairs than are usually found in the other descendants of 'Irish Birdcatcher.'

### 'WILD DAYRELL.'

A brown horse, thirteen years old, by 'Ion;' dam 'Ellen Middleton,' by 'Bay Middleton;' grandam 'Myrrha,' by 'Malek;' great-grandam 'Bessy,' by 'Young Gouty,' 'Grandiflora,' by 'Sir Harry Dimsdale.'

To describe the general outline and shapes of this magnificent animal is a task which, I candidly confess, has been better accomplished by some of those gentlemen who have already offered their useful, and unquestionably ex-perienced hints, to the admirers of horseflesh. In common with most people, I always like to see a fine specimen of the racehorse ; and do not hesitate to state that I never beheld the superior of 'Wild Dayrell,' as far as his out-line and general formation — (although there are others with more muscular development and greater -power) — putting aside his performances *in toto*. Take him, as he walked beside the 1600-guineas 'Jack Sheppard' (who was purchased to lead him to work) on the morning of the Derby on Epsom Downs. A casual passer-by would fancy he was looking at a horse and a pony ; there was as much difference in their length and size as between a railway-train returning from a race-meeting and a donkey's cart. Poor 'Jack Shepherd,' as he walked beside his magnifi-cent companion, had all the appearance of having been

well kept to his work; and certainly bore evidence against
the supposition, " Jack's as good as his master." ' Wild
Dayrell,' with his grand, lengthy walk and stride, showed
an almost indescribable superiority in comparison to
poor Jack, who looked as if he had been carrying his
namesake, and hunted to death for a month by a troop of
dragoons.

The contrast in his size, shape, and action, even with
those of his opponent, ' Lord of the Isles,' was very peculiar; the former being of that grand, unequalled length,
stride, and sweeping style; while the latter is, in point of
fact, although a true-shaped, muscular, and well-knit
racehorse, of a totally different stamp: in action, likewise,
very dissimilar. On the morning of the race, as already
mentioned, the two opponents reminded one of a grey-
hound and a hare; so much so, that previous to the race
I told a party, who informed me that he stood to win a
large sum on ' Lord of the Isles,' "that he had as much
chance of beating ' Wild Dayrell' on that day, as he (the
backer) had of being Lord-Lieutenant of Ireland." The
fact was, ' Lord of the Isles' could hardly move; having
been evidently suffering from the effects of sore shins;
and had not recovered his severe race for the 2000 guineas
with ' St. Hubert.' To compare any horse of the present
day, as far as outline and general racing appearance, to
' Wild Dayrell,' is, with great deference to those who
differ, a mistake. If he could be improved upon (and I
believe no horse ever was foaled that could not be), it
might be as to the formation of his hocks and hind-
quarters, as to strength and position, in proportion
to his frame: but animals of his great length and
general outline are seldom so well " turned under,"

in that respect, as those compact and moderate-sized horses; such as the 'Touchstones,' 'Birdcatchers,' and 'Sweetmeats.'

'Wild Dayrell' bears, in many respects, a striking resemblance to that fine sample of the racehorse, 'Bay Middleton,' his grandsire on the dam's side; and if the former had the hind-quarters and form of 'Leamington,' in that respect he would be perfect.

The fine mixtures of various strains of blood, which flow in his veins, cannot be surpassed; and contrast strangely with the in-and-in system of the present day; his size bearing proof, to a great extent, that the latter course tends to diminish the powers and size: an opinion which has heretofore been entertained by many, although hardly borne out by specimens in the present day. As to 'Wild Dayrell's' success for the Derby, the fact is he won, and could have won, when, where, and how he liked. In my opinion, he would have won with 8 st. 7 lbs., the rest 7 st.; the second being a slow, game, but moderate animal.

The 'Ion' blood is most valuable, both for speed and stoutness; he was himself a first-class racehorse, having won the Clearwell and other stakes; and ran second to 'Amato' for the Derby; beating 'Grey Momus' and others: and second to 'Don John' for the Doncaster St. Leger; beating 'Lancercost' and others. His son, 'Ionian,' ran second to 'Orlando' for the Derby.

'Pelion' (his son) was a first-class horse, especially for a mile; and 'Poodle' (own brother to 'Pelion') was a very game one. 'Buccaneer' and others also prove its value. On the dam's side, the 'Bay Middleton' strain has been well tested and proved; instance 'The Flying Dutch-

man,' 'Andover,' 'Anton,' 'The Hermit' (winner of the
2000 guineas), 'Fly-by-Night' a much better horse than
generally supposed, although not on a large scale, but a long,
low, deep-girthed one; who afforded such unmistakable
proof of what a little time will accomplish in condition:
his running in the Derby and at Ascot within three weeks,
to wit. He showed great speed up the Choking Hill on
Epsom Downs, when 'Bartholomew,' in black, appeared
one hundred yards ahead, although as far behind at the
finish. Had the Derby been run during Ascot week,
there would have been few, if any, before him.

'Wild Dayrell' has produced several winners, as set
forth — 'Avalanche,' 'Hurricane,' 'Buccaneer,' 'Horror,'
'Dusk,' 'Tornado,' 'Wild Agnes,' and several others;
besides some reported flyers in the back-ground. Still
this horse, considering the chances he has had, has not, so
far, *proved a nonpareil.*

### 'WINDHOUND.'

A brown horse, eighteen years old, by 'Pantaloon;'
dam 'Phryne,' by 'Touchstone;' grandam 'Decoy,' by
'Filho da Puta;' great-grandam 'Finesse,' by 'Peruvian.'

There can hardly be a doubt as to this horse's being
the sire of 'Thormanby,' for the reasons before mentioned.
His relationship to that running family, 'Hobbie Noble,'
'The Reiver,' and 'Elthiron' (his own brothers), as also
the fact of his being a son of 'Pantaloon,' and his dam
by 'Touchstone,' should recommend him to the notice of
breeders. He never started, having met with an accident,
but was tried and believed to be a good horse.

### 'YELLOW JACK.'

A chestnut horse, twelve years old, by ' Irish Bird-catcher ;' dam 'Jamaica,' by 'Liverpool,' out of ' Preserve.'

There can hardly be a greater proof of the " glorious uncertainty" of the turf than that furnished upon reference to the career of this horse, probably unexampled for disappointments, he having run second for the Derby, second for the Chester cup, besides his other engagements. Had he been " first" upon these occasions, what an extraordinary difference it would have made ! not only as to the amount which would have been won by stakes, &c., but as to his celebrity subsequently as a sire.

It seems strange, how frequently breeders become prejudiced in favour of *absolute winners* of such races, disregarding *in toto* the merits of " *seconds*," and overlooking many valuable qualities in which the latter, in numerous respects, even excel their victors. Here is an animal, comparatively speaking, unpatronised, because he did not absolutely win all his engagements ; and at the same time he possesses many good points as to shape, as well as most fashionable strains of blood, which should entitle him to the notice of breeders. He is of good size, length, and very racing-like in his general contour. He could stay beyond question, and was, although unfortunately notorious for " seconds," a very game horse ; as was also ' Cariboo,' his half-brother, who could run for a month.

It is, however, impossible that many valuable and promising sires, in the present day, can have a fair chance, the country being overrun with indifferent stallions and bad judges, who look to pence and throw away pounds. ' Yellow Jack' cost when a yearling, at auction, one thousand guineas, and looked as well worth the money as any yearling ever sold.

## 'Zuyder Zee.'

A dark-bay horse, eleven years old, by 'Orlando;' dam 'Barbelle' (dam of 'Flying Dutchman,' 'Van Tromp,' &c.), by 'Sandbeck.'

Probably few sires are more deserving of notice than this one; and, although the last in the alphabet, is by no means less worthy of the support of breeders, or less likely to become a first-class sire; for although he was not blessed with the best of tempers, still it can in him hardly be looked upon as natural to his family on either side — quite the contrary; and may have had its origin in very trifling, although common circumstances. Nor is such a drawback so hereditary in the sire as in the dam. That he was a first-class racehorse cannot be denied; that he comes from running families, both on the side of sire and dam, is equally true. In shape he is a remarkably handsome racehorse all over, with good size, length, and substance; and was an exceedingly sound animal, training on and lasting, with good clean legs: his colour a beautiful rich dark bay. His performances were of the first class, having beaten, amongst other good horses, and carrying heavy weights, 'Saunterer,' 'Gemma di Vergy,' &c. He won the Chesterfield cup at Goodwood, carrying the top weight; the Fitzwilliam stakes at Doncaster, carrying 9 st. 6 lbs.; the Granby handicap, carrying 11 st. 4 lbs., &c.; and ended his racing career, at six years old, as few others ever do, *perfectly sound*.

There can be no possible reason why 'Zuyder Zee'

should not prove a first-class stallion ; and, to my mind, few others are more deserving of patronage.

The following would appear the best representatives of their tried sires, and most likely to distinguish themselves at stud : —

| | |
|---|---|
| Chanticleer | Vengeance. |
| Ethelbert | Big Ben. |
| Faugh-a-Ballagh | Leamington. |
| Flying Dutchman | Ellington and Amsterdam. |
| Irish Birdcatcher | Yellow Jack. |
| Kingston | Caractacus. |
| Longbow | Toxophilite. |
| Lord of the Isles | Dundee. |
| Melbourne | Oulston, Prime Minister, and Cannobie. |
| Nabob | Nutbourne. |
| Newminster | Lord Clifden. |
| Orlando | Crater, Fazzoletto, and Zuyder Zee. |
| Rataplan | Kettledrum. |
| Sir Hercules | Gunboat, Lifeboat, and Gemma di Vergy. |
| Stockwell | Asteroid, Blair Athol, Marquis, St. Albans, and Thunderbolt. |
| Sweetmeat | Macaroni and Sweetsauce. |
| Van Tromp | Van Galen. |
| Voltigeur | Vedette and Cavendish. |
| Weatherbit | Beadsman. |
| Wild Dayrell | Buccaneer and Horror. |

In addition to the above, there are several young stallions more likely to prove successful if they get a fair chance, than they are to obtain that chance; amongst others, ' Adamas,' ' Cannobie,' ' Drogheda,' ' Horror,' ' Marionette,' ' Marsyas,' ' M.D.,' ' Mainstone,' ' Sedbury ' (a nice horse), ' Sugar-Plum,' ' Vengeance,' by ' Chanticleer ' (one of the nicest and best horses of the lot), and 'Lambourne,' a *beau idéal* of a racehorse, if on a larger scale; but he is *multum in parvo*—all muscle, and a "beauty."

Having glanced over a few sires which most take my fancy, and without drawing any invidious comparisons —on the contrary, admitting that there may be many others equally desirable—I have merely to add that, according to the official returns, there are not more than about 2200 thoroughbred mares at stud at present; those statistics being compiled with great care and labour: the number of sires being about 300—some of them having their subscriptions full. Then, how can the others possibly have a fair chance, and how can they pay? To this fact may be traced the causes of now and then finding a really good animal set down as the produce of an *unfashionable* sire. Why unfashionable? Because he has not done impossibilities—got racehorses without the chance of doing so!

It really seems strange that people who take such a deep interest in breeding should confine their attention and remarks almost exclusively to the merits of the *sire*, in many cases totally disregarding those of the *dam*. How often are wretched brutes sent to valuable sires! And what is the consequence? The owner of the sire has frequently the gratification (?) of hearing,—" Oh, there is a pretty specimen of So-and-So's stock ! " " The cross does not suit ! " &c. I know a party who is at this moment hiring stallions, and, in my opinion, he has as good of his own, if not better!

To what are the causes of failure frequently attributable, although the crossing, the value of animals, and in a great measure the lottery of breeding, are overlooked?

Some are half-fed; some half-starved; some stall-fed, like oxen, and never exercised, although called upon at so early an age to display their agility and freedom of action. It is positively amusing to behold the specimens that are bred by some persons; yet if one were candidly to give his opinion to the proprietors, he would seriously jeopardise his chance of an invitation to dinner: for few like to hear their *horse* abused, although many remain silent listeners to the slander of their absent friends by the tongue of the dastardly maligner.

The fact is, that horse-breeding, like horse-racing, is to a great extent a game of chance. But how in reason can a stallion, getting perhaps half-a-dozen mares (half the number being brutes), be expected to produce as many good animals as those fashionable ones (some of the latter, taking their chances into consideration, perfect impostors)? Then, again, the "rage" is all after certain picked ones; and during the temporary "mania," which recent success on the turf may have caused, in favour of a young beginner, he may be deserted because he did not at once prove successful at stud! Take, for instance, 'Marsyas,' the vanquisher of 'King Tom' at two years old; where is there a nicer-bred horse? Was he not a racehorse? His stock are very large and powerful; for example, 'Money-Spinner,' out of a mare that I sold to the proprietor of the magnificent monster stud at Middle Park; 'Calcavella,' by 'Irish Birdcatcher,' out of 'Burgundy's' dam; for although she did not produce well to other sires, still, owing to the patience of the breeder of 'Caractacus,' she, as well as 'Marsyas,' furnished one that could run.

Amongst the sires of the day we have some very

fine specimens, yet differing wonderfully in character and shape generally, each breed resembling the other in their particular peculiarities and qualities: for instance, 'Stockwell,' 'Rataplan,' 'St. Albans,' 'Thunderbolt,' 'Kettledrum,' 'The Marquis,' and many others, descendants of 'The Baron,' are remarkable for their immense fine frame and substance; being, in fact, mountains of muscle, like 'Nutbourne,' who is of a similar class, as well as 'Big Ben,' and other sons of 'Ethelbert.' Then we find horses of a totally different stamp, yet by no means inferior as racehorses, being more lengthy and racing-like to the eye at first sight: there are others differing from both, being the really true-made racehorse, of medium size, with great symmetry, fine racing points, level made, not possessing too much power in one respect and deficient therein in others, rendering the former more injurious than beneficial. 'Lord of the Isles,' 'Newminster,' 'Voltigeur,' 'Vedette,' 'Beadsman,' 'Caractacus,' 'Cavendish,' 'Crater,' and others, come under this class; and again, of the lengthy, slashing, racehorse style, 'Wild Dayrell,' 'Leamington,' 'Thormanby,' 'Buccaneer,' and 'Dundee.' 'Toxophilite,' son of the splendid 'Longbow,' is one of those, like 'Fazzoletto,' and others—great, large animals, presenting the appearance of being top-heavy and unwieldy, demonstrating the fact "that they run in all shapes." Still, I incline to think they are *more frequently* racehorses, and stay longest, and that their legs last and wear better, when they are of medium size—"long and low."

Some people purchase racehorses like victuallers in an oxen market, by weight. With respect to crosses, and the arriving at conclusions that particular ones suit, I ap-

prehend that the proper way to test that point is to take
the cases where the *first-class* racehorse has been the produce ; for, in my humble opinion, if all the crack sires of
the present day and all the brood mares were turned loose
into a park, and the produce taken up and trained,
amongst the number would be found plenty as good as
all the skill of some breeders has furnished : for really
their failures, taking into consideration the number of
animals they breed, are marvellous.   Any boy who has
learned the rule of three must know, that if a mare by
' Sweetmeat' has produced by a son of ' Touchstone' the
best horse of his year, that she is, consequently, likely to
produce another good one by the same sire; still it does
not follow that, because that produce was the best of his
day, he might not be improved upon : for it might be
like the little schoolboy, who informed his father that he
was "third" in his class, but, unfortunately for the
father's pride, it turned out that there were only "three"
in it.*

As to in-and-in breeding, every day proves that it is
moonshine to object to it, until somebody furnishes proofs
against it.   As to size and staying, let us look at the
best performers of the day; amongst others ' Asteroid,'

* Upon the subject of turning stallions and mares into a field,
a curious experiment was tried by myself and a friend in the
following manner, and in consequence of the following scene :—
A neighbour of mine, who was owner of a stallion, happened to
witness ' Mountain Deer ' (when he had just been put to stud, and
was very difficult to command), break loose from his groom, and the
men in charge of the mare let her head free, which caused a scene
that appeared likely to terminate in certain injury to either, or
both, if not prove fatal to one.   Upon speaking of it, I agreed, after
some difference on the subject, to try the experiment, he providing
the sire and I the mare ; and the result plainly proved that

one of the very finest specimens, and one that can run any distance.*

Shape is the principal point to have regard to in order to amend faults of sire or dam; and almost every person who likes a horse, and has any experience, knows what good shapes are. The grand secret is to know when they are *properly put together,* and to discover where the screw is loose in the machinery which renders the whole useless. In my opinion, many persons fail in breeding and amending faults in large mares, by breeding from overlarge sires, or *vice versâ;* in fact, by endeavouring to amend defects on either side by having recourse to too many counterbalancing requisites.

Having thus endeavoured to place before the reader certain remarks, which I trust may prove of some service at least, and hoping he will make allowance for any errors which may have crept in, or delusions under which the writer may labour—a failing to which all men are more or less liable—I have only to conclude these notes by trusting that they have been rendered in at least an intelligible manner, and that the advice of the poet has been taken,—

> " In fine, to whatsoever you aspire,
> Let it be simple and entire."

---

nothing whatever occurred more than if they had been reared together. Both were turned out loose into a large field, although the sire had been stabled, and they were not permitted even to *see* each other previously, the mare being left far off in the field, the horse (a racehorse) turned perfectly loose. The accident referred to was the more likely to end injuriously, the mare being " hobbled," and having fallen down, as well as the sire—somewhat resembling a Spanish bull-fight.

 * If some of those enterprising breeders should feel disposed to improve upon the in-and-in system, &c., they might succeed, like Christopher Columbus.

# BROOD MARES.

"Oft expectation fails, and most oft there
Where most it promises; and oft it hits
Where hope is coldest, and despair most sits."

THERE can be little doubt that the object of all classes following the pursuit of horseracing or breeding, is to bring to perfection, as nearly as possible, their efforts to produce the best animals. The question, therefore, for consideration is simply, how they are to accomplish it? The brood mare is the foundation upon which success principally depends—the fountain from which it must flow; it therefore becomes necessary to observe every caution in order to carry out the wishes of the breeder, and bearing in mind the very great competition in the present day, as also the very remunerative prices paid for first-class yearlings, &c., the breeder with capital should not hesitate to invest in the best animals possible (and if he have not capital, leave it alone), for the expenses of keep are quite as heavy, no matter what their quality or value may be; and as the country is overrun with moderately classed ones, they hardly can be expected to pay, for, taking into consideration the chances of missing and other losses, exclusive of the regular expenses attending

them, it really requires a yearling to realise a pretty round
price to the owner to make amends for the disappointments and expenses attending breeding. At the same
time the prices sometimes paid for *untried* brood mares
(because they are fashionably bred, and have proved winners) are quite absurd ; the long purse frequently taking
the place of practical knowledge or real judgment : for
there are some who follow this pursuit who will not be
instructed, through prejudice or obstinacy, more sometimes through the absence of natural taste or judgment,
verifying the fact that,—

> "Some men in life assume a part
>　　For which no talent they possess,
> Yet wonder that, with all their art,
>　　They meet no better with success."

Then my advice to a beginner is to select the brood
mare from the most fashionable, and, of all things, the most
*running families*, with constitution, shapes, youth, temper,
and speed. The question then is—How is a purchaser to
select a brood mare ? That query is answered thus,—Deal
or try where you will, at best it is a lottery ; but in order
to reduce the risk as far as possible, the reader should
adopt the following course :—

The *running blood* on both sides ; and there we find
them in all shapes. Some are prejudiced in favour of large
mares (generally termed " roomy " mares), and the idea is
right to a certain extent ; but, assuming that the owner is
desirous to breed a " racehorse," my opinions are hereafter
conveyed as to the sort of mare from which he should elect
to breed. Tall mares are not the more desirable *because
they are tall :* as a general rule, the deep-girdled, large-

bodied, short-legged mare, with wide hips and length, of moderate height — say fifteen hands and a half (many first-class and tried mares have not exceeded fifteen hands), if anything resembling, when in stud form, more the draught mare than the light thorough-bred — is the sort to breed from : for instance, a better illustration could hardly be afforded than old ' Echidna,' dam of ' The Baron,' (sire of ' Stockwell' and ' Rataplan'), who was more like an animal that had been drawing a float or an omnibus all her life, than breeding St. Leger winners, as she walked about the paddocks at Jockey Hall, with a head like a fiddle-case, with room for the bow on each side in the shape of a pair of ears, which her owner was so wont to explain as extraordinary and peculiar to her family, as to the manner in which they were set on, a peculiarity best seen when standing exactly in front of her descendants ; together with the prominent forehead so apparent in ' Stockwell,' &c. as in his sire. There was ' Echidna,' the daughter of ' Economist,' the dam of ' The Baron,' and his own brother ' Bandy,' who afforded so curious a proof of the freaks of nature — foaled a cripple, without the use of his hocks, literally resting on the ground, like a hare in her form, and about to be destroyed, yet grew up, with time and strength, until the malformation almost entirely disappeared, and he subsequently proved the sire of race-horses. And why not ? It was not hereditary; it was simply " a freak of nature !" One might as well argue, that because the mare exhibited some thirty years ago at Donnybrook fair and elsewhere had *eight legs*, all her produce should, as a natural consequence, have the same number. There was never a greater mistake than to suppose that breeding, no matter how scientifically carried out, is not a lottery; still, much depends upon many incidental circumstances, which are frequently taken no notice of.

Before digressing from the subject of shape, I wish to refer to a stamp of mare (before partly referred to), of which I am particularly fond, and recommend the reader not to disregard, viz. the short-legged, moderately-sized animal, as to height, &c.; with good shoulders and plenty of length, and otherwise possessing the necessary shapes of the racehorse (elsewhere described), especially avoiding a short neck, which I detest in any horse or mare. From the form described many first-class animals have been bred, and it has only to be tried to be proved to the satisfaction of any dubious breeder; for, to my mind, want of average size is frequent on the part of the sire. In proof of which I could mention many cases, having seen more of the finest horses (certainly the most level and racing-like), not only in class as racehorses, but with good size, the sons of such mares. Most assuredly the great, tall, weak-leggy animal, seldom if ever comes from the mare described. Whereas a mare of moderate size, of say fifteen and an inch, with substance, will produce by a stallion of say sixteen hands, an animal as to size a medium between the two, without the top-heavy appearance, &c. In my opinion, many mistakes are made in breeding from those over-fine or over-large mares, with very large sires. The produce may be extremely large in proportion, still we seldom see those horses over-good, finish, stay, or wear as long as the other stamp. They are generally top-heavy, and finish "like a ship in a storm." It is said, "a good big one will beat a good little one." True: but how many are there in proportion?

The next necessary qualification calling for the attention of the breeder is "temper;" a recommendation essentially requisite, for there is hardly one failing more hereditary. We seldom find mares that have been naturally bad-tempered or fretful that were not, to use a racing term, "soft-hearted jades" during their career on the turf—a drawback which their produce too frequently inherit. No matter how game the sire may be, his reputation frequently becomes injured through the produce showing the softness of the dam. With regard to temper, I have observed that bad-tempered mares are more frequently chestnut than of any other colour.

A difference of opinion exists as to whether the produce takes, as a general rule, more after the sire or the dam; many persons believe the latter to be the case: in which opinion I concur, especially where the constitution of the mare is strong and unimpaired, and has not been affected by disease, heavy distemper, or over-training: that is to say, the produce, as to perfections or imperfections in shape, colour, and temper, as well as the other qualities, will, in the majority of instances, more resemble the dam than the sire; yet not unfrequently, through freaks of nature, will bear a much greater resemblance to some of their ancestors, as far removed, perhaps, as two or three generations: for instance, as to colour, we frequently find a black colt or filly by a chestnut or bay sire and dam: in which case the anxious owner diligently seeks and traces back the pedigree, being naturally desirous to ascertain "where the colour" comes from; and he, no doubt, believes the produce "takes after" the most "distinguished" of his ancestors. For instance, one would naturally assume that 'Saunterer,' a black horse, inherited his colour from 'Sir Hercules' (his grandsire). Another instance is 'Thormanby' (whose sires, 'Melbourne' or 'Windhound;' the former was bay, the latter brown; his dam, 'Alice Hawthorne,' bay): one would fancy he took his colour from 'Pantaloon,' sire of 'Windhound;' a horse of the very same colour.

The most important question to be considered, and, as far as possible, to be solved, is, how far that admitted fact, that "like begets like," holds good in breeding racehorses; and what perfections and imperfections are most hereditary? Let us take, for example, the cases of the "mule" and the "jennet." The former, as everybody knows, is the produce of a mare by an ass; the latter, that of a mare-ass by a horse, or pony. Then, do they not bear testimony, to a great extent, in favour of the argument, that the produce, in the usual course of horse-breeding, must follow more the qualities of the dam? for it can hardly be argued that the colour of the mule, which, in ninety-nine cases out of one hundred, is brown, or the darkest bay, is not that of the dam: the dark-brown ass being seldom seen, in proportion to the other. Then as to the jennet, it has always more of the black streaks or stripes on the arms and legs, and along the back, than the mule—a further illustration and proof in favour of the dam; the colour invariably being of a lighter hue, more approaching that of the ass.

Then, again, as to size; who ever saw a jennet as large as a mule? the latter being occasionally as large as a horse. It is equally true that the jennet is invariably got by a small pony. Moreover, we have seen even racehorses with the dark streak along the back, when the colour is bay, either light or dark, approaching that of the jennet.

And where we find a sort of cream, mealy, bay colour, those streaks become more numerous and marked, upon the arms especially, resembling the zebra. I had one, many years ago; and, curiously enough, he was not only thus marked, but appeared to have a most peculiar temper, resembling in his ways and acts the mule in many respects, with wonderful endurance: and although he was marked exactly like, and of the precise colour of the jennet, still he was an average-sized horse, of about fifteen-and-a-half hands high. It has been stated that the produce of a mare, having previously had produce by an ass, has been known to have borne the marks and streaks referred to for several seasons subsequently, which have gradually died out. There is one thing quite certain, we frequently find even racehorses with the black streak along their back, even to the very root of the tail: ' Wild Huntsman,' by ' Harkaway,' for instance — (not that I mean to insinuate anything derogatory to 'The Huntsman,' who was a very good horse, indeed)—and there are many others besides with this peculiarity.

It a well-known and proved fact, that there is nothing like the *tried* brood mare, of fashionable running family; still, admitting the fact, it is not so easy to become possessed of them without paying, in many instances, exorbitant prices: therefore, a beginner would do well to secure some of the descendants, say daughters, of such mares, provided they have no drawback and are sound, and got by horses of running and fashionable blood. For it is truly astonishing how frequently owners put such valuable mares to brutes of no pretensions either to first-class strains or running family. The " poison " thus sown does more mischief than at first imagined, and it requires generations of superior crossing to eradicate it. As to shape and size, as before remarked, plenty of length, with strength combined, about fifteen-and-a-half hands high, neither too short nor too long in the leg, with good length of "*arm*," and *muscular*, although appearing to the eye at first sight, and when in stud form, rather shorter in the leg than otherwise; with good, clean sinews, and sound, well-formed feet: for there is nothing more fatal nor hereditary than small, contracted feet, which render the finest animal, in other respects, worthless.*

.Although the plain-looking mare of plenty of substance is the one to choose, still, with the "plain" appearance through the frame, the quality will be found in the head, neck, and shoulders of the well-bred mare; although some of the best blood, when at stud, present the appearance of common draught mares, being a peculiarity to their respective breeds. And I confess that I, for one, am not an admirer of those over-pretty-headed animals for pecuniary purposes, however useful and desirable they may be for Rotten Row. 'Teddington' was the prettiest-headed horse I ever saw for a good one. I have seen few "pony-headed horses" of the first class. Give me a

---

* A friend of mine had bred for seven years from a very fine and well-bred mare, and had put her to several different sires, and yet the produce were all literally useless for any purpose. In speaking with him on the subject, and offering to purchase her, he accepted of fifty pounds as her price. I had not seen her during her career at stud, but when delivered she was in a wretched condition—a perfect skeleton. Having given her every care during the winter, and put her to 'Mountain Deer,' she produced the following years two colts, both of which turned out very good racehorses, the dam herself growing into a splendid mare, her former owner absolutely not knowing her the following year.

clean, good, bony head, of fair size and average beauty; with a sensible, steady eye, clear and bright, and not flighty; for nothing shows signs of the temper more than the eye, especially in mares. Although we sometimes find those wiry, light mares, successful at stud, still, as a general rule, the others are the sort, especially to breed stout stock. A tall, leggy, overgrown mare, is by no means the class of animal to breed from : for their stock invariably, though possessing height, and, at first sight, of commanding appearance, do not represent the level, muscular, and equally-proportioned points of the true-made racehorse; but when trained, and the flesh necessarily reduced, frequently turn out top-heavy, weak, and worthless, not possessing the stamina of animals bred from the class of mares before described. As an instance of the success at stud of mares of the stamp referred to, ' Clari ' was a most striking one. She was the property of the late Mr. Watts, who bred 'The Baron,' and other good horses. This little mare produced by ' Magpie ' (son of ' Young Blacklock '), ' Chat,' ' Chatterer,' ' Chit-Chat,' ' Chatterbox,' and ' Third of May,' and others, all of fine size, and good runners. She resembled more a little hack-mare than the dam of racehorses.

However prejudiced the writer may be in favour of the moderate-sized mare as a general rule, he by no means denies that the produce of a large mare, provided she be equally proportioned, is not more desirable, and in certain instances more likely to turn out " first-raters," the produce of such mares being invariably very good or very moderate. These remarks merely refer to the class of mare which produces most frequently good stock.

With regard to the mare that has proved herself of the first class during her racing career, let us contrast the probable success of her produce, and, for argument sake, take any of the first class of the present day that have so distinguished themselves; put those mares to the crack sires of the day : then, on the other hand, select an equal number of the daughters of such mares, assuming that they are fresh, sound, and of equally fashionable and running blood on their sires' side, and possessing the average shapes, &c., yet, through accident or other causes, may not have ever started, or perhaps ever have been trained.    What would be the chances against the produce of the young ones as a lot, beating those of the old, by the same sires, taking for granted that there was no drawback as to breeding and shapes on the side of the young mares, and assuming that their dams had not previously produced winners ?    I, for one, would select the produce of the young ones in preference ; and why ?    Because we have innumerable instances where mares that have been the best of their day as racehorses, have proved the most signal failures at stud; and, at the same time, we have seen the best runners the produce of mares of no note as racehorses.    And why ?    Because the constitution and system of the old and valued servant has been too frequently impaired ;    and if she does produce a son or daughter worthy of her, it does not frequently happen until she has had considerable time to recover her lost natural vigour of constitution, which seldom entirely returns : yet she hands down to her offspring, in the young fresh mare, her racing qualities, as far at least as blood, shape, &c., unimpaired by excessive exercise or over-drawn development necessarily attending her racing career.    Take old

'Beeswing,' for instance; one would have expected the best vintage from her alliance with 'Sir Hercules,' yet, although the produce was 'Old Port,' it was not good; as also 'Lord Fauconberg,' by 'Irish Birdcatcher,' out of 'Alice Hawthorne.' Then, of more recent date, we find those magnificent mares, 'Maid of Masham,' 'Virago,' 'Lady Evelyn,' and many others, comparative failures; next that extraordinary mare 'Crucifix,' although she did produce one first-class horse, 'Surplice,' still her other produce being so very inferior, with the exception of 'Cowl,' one could hardly arrive at any other conclusion than that, to a very great extent, it had its origin in the exhaustion of the constitution. Suppose that she or any other of the dams of celebrated horses, having, like her, done so much work upon the turf, had, from accident or other causes, never been trained at all; there cannot be a shadow of doubt that the produce would have been even superior, and the dams would have lasted longer, and produced more successful stock.

Moreover, one of the reasons why more winners are descended from those mares of celebrity on the turf than many other untried ones, is this; that owners more frequently give them every chance than they do to others less celebrated, although probably better adapted for the purposes of breeding. How many valuable mares have never had any chance? I have, on many occasions, even in the Hansom cabs of London, been struck with the shapes, qualities, and other recommendations of mares, and in some instances taken the trouble to ascertain all particulars relating to their antecedents, when they have turned out to be some of the best blood in England. If I were asked to-morrow to select one hundred mares (not

T

confining the order to thorough-bred alone, but for general purposes to improve the breed of *useful* horses), the
post I should select would be the thoroughfares of London. I have seen some extraordinarily well-shaped and
useful mares in those cabs. The idea may appear far-
fetched; yet the fact is so. On the subject of constitution,
I would simply recommend the breeder cautiously to avoid
any mare that does not possess a thoroughly sound and
unimpaired one; for let their racing merit have been
ever so great, they are literally useless for breeding if
they are not thoroughly sound in this respect. Many
a time I have remarked, that the dams of "really
good horses" feed much better than others (one of
the best proofs of sound constitution in any animal);
indeed I have known some that appeared never to cease,
and moreover, showed it; which is not always the case.
On the other hand, there are mares naturally of fine frame
and substance; in short, everything to look at that
the most fastidious judge could wish for—apparently
full of flesh and vigour—and yet they are in reality
impaired in constitution; and although their produce
may, like themselves, have an outward show of health
and condition (although not good feeders), you will
find them either diminutive in size and substance, in
proportion to the dam, sire, and ancestors, or otherwise
deficient in staying powers or endurance; and this rule
will apply equally to the produce of mares or sires that
may have for years produced the finest stock, and possessed
the soundest constitutions, but, becoming worn out by age
and natural causes, have ceased to retain that power and
vigour necessary in the brood mare. Even here there are
exceptional cases, such as 'Faugh-a-Ballagh's' dam,

'Guiccioli,' who was far beyond twenty years old when she produced him; and according to the records of breeding and racing nominations, subsequently had not only produce, but produced twins—a colt and a filly—own brother and sister to 'Faugh-a-Ballagh,' (as if a halo of glory had been cast around her by the success of her renowned son), neither of which, however, appears to have proved very successful at stud or otherwise, especially the colt called 'Thanamadowl:' a curious instance of the "uncertainty of breeding." The fact of this mare having, at so late a period of life, produced twins, is a very strong argument against the doctrine which I have endeavoured to support.

As before stated, one of the many essential qualities, as well as being one most hereditary, is temper, and therefore especial regard should be had thereto; the want of it being easily discovered requires no comment, further than to assure the reader that, without this necessary qualification, any animal is not only worthless, but a perfect nuisance, and always unprofitable. Such being the case, it behoves the speculator to have regard to the fact and learn, if he does not happen personally to be in possession of it from knowledge of the mare's antecedents, that on this score she is desirable as a brood mare, having during her racing career possessed such a character: for it must be patent to any person, with even a simple knowledge of such matters, that when one invests his capital in such precarious undertakings no chance should be thrown away, inasmuch as it is like taking a lease of a house, where a tenant or purchaser cannot be at all times found; the entire value of the produce of the animal depending not only upon her own performances as a racehorse, but upon those of her produce. Then, as experience has proved (at least to those who have had it practically) that nothing is more hereditary than temper, the want of it should be strictly avoided.

When an owner has proved successful with any particular animal, his keeping to " the bridge that has carried him safely over" is, perhaps, far more deeply displayed by a repetition of the purchase of the "next of kin," than by his gratitude to his fellow-man for far greater services otherwise rendered : for it is truly amazing with what tenacity (in competition at auction, or otherwise) he adheres to a brother or sister to one that may have done good service ; consequently, in many instances absurd prices are realised, even about three times their real value—to the great, although unexpressed, delight of the breeder, who, no doubt, in many instances, has more brains than the purchasers or bidders. This proves the necessity, at least for selling purposes, of adhering to the " tried" brood mares ; for there is, as a general rule, nothing like it, although we have instances where some display a patience, when disappointed, almost rivalling that of Job, and then have succeeded : and if in this respect there is now a living representative of that patriarch, we are bound to believe he is to be found in the person of the enterprising proprietor of Middle Park Stud Farm, the breeder of ' Caractacus,' 'Queen Bertha,' &c., who appears, when purchasing, to set outlay and competition at defiance, and who really must become a second Crœsus, if he obtains for the offspring of his increasing and unsurpassed stud, prices commensurate with his expenditure and liberal views when purchasing. Upon the subject of " like begetting like," every person who breeds seeks to produce (if possible) a facsimile of the sire or dam. Such being his attempt, then the question is how to do so, and how far his expectations are likely to be realised? Let him remember that

" Birds breed not vipers, tigers nurse not lambs."

The idea of the uncertainty of breeding has, to a very great extent, its origin in the following facts : In the first place, many men make the attempt without the slightest taste, much less knowledge, practical or otherwise, of the animal ; they consult others, who are equally ignorant in such matters, and breed from bad animals. Secondly, others, who know, perhaps, very little more, and pretend to a great deal more, are too fond of their money to send to the best sires, or select the best mares. Lastly, there are those who have both mares and means, and, as far as they are concerned, would spare no expense, but who having still to know how such animals should be attended to, are in very many instances deceived : for it is truly incredible to what an extent they are neglected, even at the time they appear to ordinary observation to look well and in good condition, as most observers would fancy ; yet, to a judge, the stamina upon " handling" is not there. The crest or neck of a brood mare, and her condition otherwise, to insure the " tip-top" produce, should be not only in a fleshy state, but, comparatively speaking, as firm as if in training, although proportionably treble as to thickness. Another proof of the well-fed mare will be seen about the end of March or beginning of April, when the long old winter coat not only commences to, but rapidly falls off, in a sort of wool (if the term may be applied), much darker or lighter, according to the colour of the mare, than the fresh coat; when here and there, in patches, the short, bright coat of summer will present itself, indicative that the soil underneath, in the shape of good food, has been well tended, and that the crop is accordingly early. The old hair, with one stroke of the hand, will come from a well-fed mare more freely than with a curry-comb and brush from a dozen half-fed ones, and months earlier. In a few words, the time to lay the foundation for the future of the foal is while the dam is carrying it; or, in Irish vernacular, to make him a racehorse "you must do so before he is born," because in racing, in all countries, they contend when very young, and cannot afford to throw any chance away.

Many brood mares are neglected as to *health*, some persons believing that, as long as they are on good pasture, &c., that is sufficient; whereas it frequently happens that they are in bad health through some inward ailment, which medicine could remedy. We frequently see mares on the best pasture, still they do not thrive and are completely without flesh, and "staring" in their coats. There is no medicine so successful, yet harmless, as linseed oil.

With regard to soundness in brood mares, admitting that to have it to perfection would be most desirable, as in every other case, still many valuable mares are rejected for mere trifling blemishes or nominal unsoundness. It may astonish many admirers of horseflesh to learn, that many of the best-tried brood mares have been spavined, blind, and otherwise unsound, and yet not one of their produce inherited their disease; and why? Because, in most instances, admitting the many diseases to which horseflesh is heir, more of them are the effects of accidents, ill-treatment, or want of proper attention, than natural causes. Amongst others, the dam of 'Burgundy' was spavined on both legs, yet none more sound than her produce. I do not mean to recommend the animal with even an eyesore, in preference to, or as desirable as those without it, but most decidedly to maintain, that there are innumerable instances where we shall find most extraordinary exceptions to the rule of perfection in shapes, or soundness, in favour of mares that, to the inexperienced eye, present the appearance of cart-mares or hacks, with various blemishes, apparently lessening their value; yet, in reality, such mares, for breeding purposes, may be quite as valuable, although in many instances purchased for merely nominal prices.

It is extraordinary how some sires have a fancy for certain mares in preference to others. I have known instances where they have refused to serve some; and although the mares have been disguised over and over again in various ways, still, after the usual signs, &c., they have returned to their manger. On the other hand, I have known in other cases, where it required double force to restrain the sire on the appearance of certain mares.

Some short time since a correspondence appeared in the public sporting journals upon the subject of certain artificial means of insuring the produce of brood mares. The operation is as old as "Kate Kearney's cat," and has proved a perfectly true and successful one. The practice of firing off a pistol immediately after service is likewise so. It is equally true that it can be easily told if a mare be with foal when half gone. I have had both tried with perfect success, and no possible danger with

either. The former is especially useful with young barren mares, or those that have missed a season; and
many of the former missings are from not adopting the
practice, which, however, is better let alone, if possible,
as there is nothing like leaving everything to Nature.
But, of all things, an *experienced* "*hand*" is required,
otherwise there might be dangerous results. Another
"old fashion" is, that of turning the mare immediately,
within view of the sire—probably after the daguerreotype
principle. A friend of mine sent a mare that had missed
for several years to other sires to a horse of mine, with
orders that no *artificial* means should be resorted to.
The lad who brought the mare was sent on an errand pur-
posely, and both the pistol and other operation resorted
to : the mare had produce. The owner subsequently declared against the system as useless, until I undeceived
him upon the point. The fact is, there can be no doubt
upon the subject.

Upon reference to the statistics of brood mares, it
would appear that there are about two thousand annually
returning the following results :—On an average one
year with the other, about one-third of the number
"miss," or are "barren;" and of the produce about one
hundred die as foals. In the first instance, the operation referred to would have the effect of diminishing the
number in an extraordinary degree; and as to the death
of foals, many of those that have been lost could have
been saved if properly attended to; and when the usual
purging appeared, a little castor oil administered. The
attendance which young mares require during the period
of their service far exceeds that supposed by many persons, and to the annoyance, want of quietness, and want

of being kept from others during that period, can be traced a great deal of the failure of produce; a large number of mares, as mentioned, being frequently "bundled together," and fresh arrivals causing disturbance amongst the whole lot.

No one but those who have experienced it can have an idea of the enormous quantity of grass, hay, corn, bran, &c., consumed by brood mares, especially in winter, when with foal, or during the period they are suckling; in fact, they appear never to cease or be satisfied. And while upon the subject of the quantity, it may be as well to offer a few remarks upon the quality and nature of the food most desirable, and what has been found from practical experience best suited. In the first place, a great deal depends upon the period of the year. During the summer months, if there be plenty of sound pasture, the brood mare, with her foal, will require but a couple of feeds of oats, and occasionally a little bran swelled or moistened, daily. The barren mare will not require any oats during that period, provided she has plenty of good pasture. During the winter they should both have—the one, three feeds of oats, bran, carrots, and occasionally linseed, mixed warm, at night. The barren mare should get at least two feeds daily from the time the grass begins to become scarce — say, end of October, during which period Swedish turnips strewn over the pasture in a sound state, and *uncut*, will be found most beneficial, and very much relished by mares, as well as a fine substitute for grass, particularly after foaling, as they tend to increase the milk. Boiled barley is also a fine nourishment for mares, with linseed and oatmeal drinks, during foaling time; for which purpose a large boiler should be con-

tinually kept in use, and turnips boiled and mixed occasionally with bran and oats, linseed meal, &c., all of
which are most useful—a change in diet being desirable. The mares should be served according as they
appear to relish any particular food. The stud-groom
should be careful as to weather and the period of service;
for in severe weather—snow or frost, for instance—
there is nothing better at night than a good warm mash
of bran and oats, and various other drinks (during the
time of foaling, and previous and subsequent thereto),
such as linseed, oatmeal, &c., given barely warm, having
been previously left steeping from a boiling state.

Attention should be paid to young mares, especially
just out of training, as they are generally dried up from
the effects of training, and are frequently almost "hide-
bound," like the bark of a tree, and not in a fit state
for stud purposes, and prove a perfect nuisance to the
proprietor of sires; in numbers of instances proving barren the first, if not the two first seasons, through not
having been properly "softened" in condition. While
upon the subject of the service of mares, I would recommend the breeder to have his mare *tried* by the
stallion she is about to be *served* by, and not by the
usual "attendant" upon crack sires—there being many
reasons which dictate the propriety of this course, as
well as that of giving exercise to the sire just previous to
service. Another mistake is frequently made, through
ignorance of the fact that young mares, when really
in season, do not, when brought near over-boisterous or
fresh young sires, or "noisy" old ones, appear so, but
display every symptom to the contrary; for, between the
causes referred to, and the usual application of the "rib-

binders," supplied by ignorant servants, with a stick or whip during the trial, the young animal becomes unmanageable.

From the first of January up to the 1st of May is the principal period which requires the attention of the stud-groom, during the time of foaling, and their service. The object should be to replace the deficiency of ordinary grass, and have, as a substitute, a field of early rye-grass, which can be preserved and *forced* during the winter and spring; and being cut with the scythe and given to the mares, will prevent its being trampled upon in the field, and otherwise wasted. A field of Swedish turnips should also form part of every breeding establishment : they are most nourishing, as well as economical, and less likely to cause diabetes; and when, as stated, strewn over the bare pasture, the mares seem to enjoy licking the particles of clay which are attached to the vegetable. I have frequently noticed them for a considerable time indulging in doing so. It may seem strange—there is no accounting for taste, although it appears extraordinary—that animals so remarkable for delicacy and fastidiousness in diet should display such a fancy; however, to a great extent they should be accommodated with what they relish, as it is natural to suppose they will thrive best upon it: although it by no means follows that man cannot improve it, for whatever they eat most of might not agree with them, no more than lampreys did with Henry I.; for an epicure or gourmand might eat himself into apoplexy. Oxen like clover, but I have known them to eat until they absolutely burst.

The corn and feeding supplied to all horses should be well cleaned and sifted, and free from sand, &c. Many animals have died from neglecting the cleansing of the corn, &c.; and when opened (as already stated) immense balls, as large as cannon-balls, have been extracted, being the cause of death, and have been preserved by veterinary surgeons, at whose establishments they are to be seen in London, resembling a piece of beautifully-grained and polished marble, and equally heavy; having been forming for years in the intestines, and being composed of sand, straw, hay, &c. The immense size and weight of those specimens are almost incredible, and make them worth inspecting as curiosities.

It is as necessary for the breeder, therefore, to secure the services of an intelligent and experienced stud-groom, as it is for the purchaser of a racehorse to provide himself with a proper trainer; and the more practical skill each can boast of in the veterinary art, the better: for it is wonderful in how many cases, and how frequently, their knowledge will be called in question, and put to the test. Fancy a breeder having one of his valuable mares, or young stock, taken suddenly ill—perhaps the dam of some Derby or St. Leger winner, or an own brother or sister to one—probably at an hour of the night, and at a dis-tance from the residence, which would render veterinary assistance almost impossible: the matter becomes serious, when perhaps thousands are at stake. At the same time many very great mistakes are made, and the results fatal, through too much interference with nature during foaling; as a most experienced and well-known veterinary surgeon of fifty years' practice (who bred some of the best horses ever foaled) informed me, "that more mares and foals were lost through interference, and not leaving nature to perform its own duty, than people had any idea of." These remarks were made upon the occasion of my having called his services in question, in a most extraordinary case, with one of my own mares, and in which the soundness of his judgment was proved; for although I believed (as did my stud-groom also) that the mare could not possibly live, she, without any interference or assistance, produced a fine foal, and both were perfectly well in the course of a few minutes. The veterinary during the time laughing at the idea of danger, and relating the opinion expressed; although he admitted it was an extraordinary case, and that, had the usual course of interference with nature been adopted, nothing could have saved the mare. Brood mares should not be disturbed, but kept quiet, and in a properly-fenced pasture, free from the annoyance of other horses; especially during the period of foaling, or while they are going through their trials : any neighing, or interruption from other strange animals, having a most injurious effect, tending to make some "pick" foul; preventing others proving with foal. One of the most dangerous nuisances about a breeding establishment is a pack of hounds, or harriers. I have known mares to gallop about for hours, and the entire stud driven into a perfectly frantic state, at the "music" of the dogs. A mare of mine, own sister to 'The Baron,' "picked" foul to 'Melbourne' through this cause; and others in the same neighbourhood suffered in a similar manner.

When mares have had their summer's run on sound, well-drained pasture, comprising not only the usual grass seeds, but a mixture of the other various seeds to which they are so partial, clover, yarrow, &c.—and which can be so much increased by occasional top-dressing; and when the period for weaning the produce arrives, great care should be taken, not only of the foal but of the dam.

As to the latter, with regard to her milk, which the foal has ceased to relieve her of; and then as to the removal of the foal, for many accidents happen, if care be not taken to have the latter well secured, and completely removed from any possibility of fraternising with the dam; as accidents frequently occur through attempts on the part of either to regain each other's company.

Before closing my remarks upon the feeding of the mares, it is right to observe that the brood mare, with a foal at foot and another coming, both draining the constitution, will require extra nourishment and stamina, even during the summer months; and oats, &c., morning and night, should be supplied. If any breeder fancies it is economy to stint a brood mare in any manner, it would be more advisable for him to leave the breeding of racehorses to others, and " amuse " himself otherwise; for he must, in the present day, either feed properly, breed properly, or leave the matter in the hands of those who do so: unless he wishes to amuse himself, get rid of his money, or swell fields of horses without a possibility of success or profit.

The proper time to wean the foal depends, to a great extent, upon the age, health, and condition, not only of the foal itself, but of the dam; for many reasons: in the first place, the dam may be old or weak, or perhaps not of a very strong constitution, and a bad nurse, so that milk becomes more injurious to the foal than beneficial: in which case it is better to wean the produce and gradually accustom it to sound diet, giving cows' milk, and other nourishing food.* The usual time allowed for suckling

* Cows' milk should be strained to avoid hairs, which have frequently formed a large ball in the intestines, and killed foals.

being from six to seven months in ordinary cases; when at weaning, the object should be to replace, as far as possible, the loss the foal naturally experiences, and by degrees train it to partake of the nourishment substituted: this, for a certain time, until accustomed to it, will be anything but relishable, more especially the cold water in lieu of the warm milk, which it is, in many instances, necessary to force the foal to drink, when suffering from excessive thirst, leaving it in a fixed reservoir in the corner of the stable, but especially avoiding (as hereafter mentioned) buckets with iron hoops, which are most dangerous; warm oatmeal drinks, mashes of bran and bruised corn, mixed frequently, given in small portions, are desirable, reducing by degrees, until the foal becomes accustomed to the usual diet. Also a few chopped carrots mixed with the corn, but not too much, as they tend to cause diabetes. Change of food is requisite, and of all things, repeatedly a soft mash of bran left soaking in boiling water, covered and mixed well, with a little bruised oats, supplied when barely warm.

When medicine is required, which is frequently known by the coat staring, and when young animals are not thriving as the owner would wish, the best possible remedy is a proper quantity of linseed oil. Some people give castor oil, which is good for very young foals, but I have *tried* and proved the *linseed* oil to be an extraordinarily efficacious remedy, when yearlings or animals eight or ten months old, or at any age, have been thus staring in their coats, dull and heavy, and not doing well. I have seen them, in an incredibly short period, with their coats laid down, and the improvement beyond description; moreover, it is a very harmless medicine: but in any case,

and at any period of the year, it is advisable to keep them
shut up during the time they are under physic, supplying
warm drinks, mashes, &c.

It is really marvellous the manner in which some
people not only stint, but absolutely starve their brood
mares, yet expect to breed racehorses. I have witnessed
(of course in silence) the miserable, unnatural practice
of such persons, some of whom were men who ought to
have known better; indeed, there can be no doubt they did,
but they remind one of the remark of the old Athenian, who
said that his countrymen knew what was right, but that the
Lacedemonians practised it. One instance I cannot easily
forget. Upon my introduction to a stud-farm I there
beheld, amongst a few others, a mare once celebrated, and,
to my mind, one that ought to be one of the best brood
mares in the world (for she and all her family could run).
There she was, a miserable spectacle to behold, especially
when one bore in mind her brilliant performances on the
racecourse — a miserable spectacle even in summer: there
was not enough of grass in the field to graze a goose;
and, as the clever and witty "Argus" said, in describing
the appearance of an animal some years ago, "The poor
mare's astonishment at the sight of a feed of oats would
best be compared to that of Robinson Crusoe when he
saw the footmark in the sand." Yet such men expect to
breed with success. The hungry, half-starved appearance
of that beautiful mare, as she anxiously approached, made
a deep impression upon me, leading me to suppose she
had been in the habit of occasionally receiving food from
some humane hand: all I can say is, if she was, neither
she nor her companions showed it; for they one and all
might well be compared to the skeleton of Jonathan Wild

on the gibbet on Clapham Common. Of course her
owner will be grievously disappointed at her not producing a Derby winner : she will "turn out a bad brood
mare." By-the-bye, the person referred to used to breed
for sale, and I believe has prudently abandoned the pursuit, as he found it did not pay. The prices realised were
amusing, although by no means encouraging to breeders
and *feeders,* in general varying from ten to fifty guineas :
the latter price being the top of the lot, which resembled
more a pack of half-starved Shetland ponies, fed upon
furze-bushes, than animals bearing the name of thoroughbred yearlings.

Upon the occasion of purchasing a yearling some
years ago (very early in the season), in very low condition,
having remonstrated with the breeder upon his want of
wisdom in having the colt in such a state, he replied,
" He had frequently sent sacks of oats to the man, the
care-taker of the farm where the colt was reared, but the
only way in which he could account for the matter was,
that the old woman kept a great number of hens." Be
that as it may, the animal made speedy improvement, and
subsequently turned out a very good horse ; of which fact
"The Racing Calendar" bears testimony.

There can be no doubt (as before stated), that the
time to begin to feed the foal is " before it is born :" by
giving the dam the best of every care, the produce invariably shows, when dropped, the attention she has received. I have seen some, when a few hours old, with
skins like satin, muscular loins, quarters, &c., and full of
spirits, jumping and kicking about as if a year old :
whereas the produce of the half-starved mare has neither
strength, condition, nor activity, being dull, staggering,

U

and weak on its legs, staring and long in its coat, with other signs indicative of the hardships or want of care experienced by the dam.

Some people fancy that the produce of a fat or high-conditioned mare is never so large as that of others. No doubt, when dropped, it may not be so large, or apparently "tall," yet the improvement in growth and muscle subsequently tells and gains gradually : in short, the produce shows in every respect where the proper care and attention have been paid as regards the dam; as an illustration of which I relate the following instance :— A mare of mine, named ' Ariadne,' by ' Irish Birdcatcher,' a particularly healthy animal, produced by 'The Mountain Deer' a beautiful foal, named ' Highland Laddie;' in about an hour after he was dropped I happened to put my hand upon his hind quarter, remarking the extraordinary muscle it presented (so frequently developed in the 'Touchstone' quarters); when putting his ears back like an old horse when being dressed over, and jumping forward with the agility of an antelope, he kicked back with both legs, with force that was really wonderful. The late Marquis of Waterford purchased this foal with his dam, but it did not prove a fortunate speculation, for although as promising as any animal in every shape (the picture of his sire), he died; and upon a post-mortem examination proved to be full of worms, which is by no means uncommon with such animals. Curiously enough, he furnished a proof in favour of the opinion expressed by many, that cows' milk tends to create worms; for he had had a quantity of it during the summer, and although he grew wonderfully in every respect, and appeared in fine health and spirits, still he did not put up flesh, and generally appeared rough and dull in his coat, which many times caused me to "blow up" the stud-groom.

Upon the subject of worms, with which foals are repeatedly troubled, a few remarks may not be misplaced. It is admitted that "prevention is better than cure;" so it becomes most necessary, the moment any symptoms present themselves, that the proper powders, which can be obtained at any veterinary establishment, should be administered, as the animal never thrives ; on the contrary, dwindles away, and frequently becomes too weak to withstand the effects of the medicine necessary to get rid of the nuisance. It has been stated, that they have their origin from the small yellow insects which so frequently cover in swarms the arms and legs (especially the former) of brood mares, and being licked off become transformed.

On the subject of feeding, I mentioned that previous to, and after foaling, a few turnips would prove beneficial; and why? Because they increase the milk (as with the cow), as well as being otherwise beneficial towards health. It has been stated by some that they have a tendency towards causing mares to "pick" or "slip" foal : such is not the case; the argument on the other side being, that oats prove a preventive. Notwithstanding such objections, my advice to the breeder is to try the system of strewing over the pasture, at the period referred to, during the day-time, sound turnips, in their usual state, uncut, which is less dangerous than when chopped or cut in pieces, and which, in the absence of grass, will be found most serviceable, and a very good and nourishing substitute at such a period : the principal object being to produce as much milk as possible for the nourishment of the produce.

Such being the case, there can be no doubt the vegetable referred to ought to be not only most approved of, but is more economical than others of the kind, such as carrots, &c. These remarks, of course, merely refer to the winter and spring season, the periods at which grass is so scarce.

With regard to the treatment of brood mares in general, during the summer season, whether barren or otherwise, taking for granted they have good pasture, with the addition to the mares with produce of oats, twice or thrice per diem according to circumstances, it is most desirable to have sheds or shelter, where they can retire from the annoyance of flies or the excessive heat of the sun at certain hours of the day, for nothing annoys them more, prevents them feeding, or otherwise interferes with their comforts. The paddock-house should, even at that period of the year, be supplied with some sweet, well-saved hay, of which the mares will partake during the heat of the day, as well as fresh-cut soil. Variety in diet proves as acceptable, beneficial, and doubtless as palatable to such animals, as it does to human beings; therefore "change and variety," to a certain extent, is desirable.

Then, as to the weaning of the foal—the care necessary to be observed as regards its dam, and the period at which it becomes necessary, to a great extent depends upon the age, as well as the strength and health of the produce; the usual time allowed for suckling (as already stated) being about six or seven months: for it must not be forgotten, in the anxiety of the owner to afford the foal at foot every chance, that the dam is perhaps carrying another, and that her constitution requires all its vigour and strength to do justice to it. At the same time, the milk which she can produce to the foal at foot reduces not only in quantity but in quality, leaving the mare's condition only sufficient to support, according to nature, the wants of the coming offspring; for, after having suckled for six or seven months, she requires time and reasonable rest to strengthen the system. The strictest attention should be observed, and every precaution taken, during the period the foal is being weaned, in order to prevent accidents, as I have already observed; the young animal, in its anxiety to rejoin the dam, will hardly hesitate at any obstacle: in proof of which, I relate a circumstance which came within my own knowledge some years ago. At the late Mr. Graydon's sale (the breeder of 'Roscius,' 'Clincher,' and others), who was a very eccentric and most extensive breeder, as well as a good judge of crossing (but an indifferent feeder), I happened to purchase two filly foals, the one own sister to 'Clincher,' the other, 'Allegrette,' grandam of 'Anfield.' Having called repeatedly the attention of the auctioneer and his attendants to the danger which might follow the removal of the dams, which had been purchased by other parties, and even taken the trouble to bring them and explain the consequences which might accrue, upon calling the next day I was informed that the first-named had jumped over the half-door, nearly four feet high, which had been left open contrary to my directions (although I had seen it locked on the evening previous); the foal, worth five hundred guineas, coming in contact with a harrow opposite the door, was killed—the spike of the harrow having entered its heart. The other filly, however, proved a very good mare; and if her grandson 'Anfield' takes after her, he will test the staying qualities of a good

many : she could " run for a week," as could her dam old
' Alba,' by ' Dandy.'

So many accidents occur to those animals, it is advisable at all times to reduce the risk as much as possible; therefore, I should recommend the reader, if his
brood mares wear head-collars, to take care that they are
of a size, and made in a shape, which will prevent the
possibility of a foal, or the mare herself, getting the
feet into them, or otherwise fastened. It is really absurd
the size we sometimes see those collars — literally about
to fall over the mare's nose. The mares sometimes
scratch their heads and ears; the foals continually jump
on the dams, pawing and playing: then, again, the
former, frequently, looking in over a gate, get fast. I
have seen them do so, and in attempting to relieve themselves pull down the gate and receive serious injury.
Such gates should not be made use of, and, of course,
would not in a properly-appointed establishment.

Some owners and managers of breeding establishments adopt a system which appears to me a most extraordinary one, viz. even in the middle of summer shutting
the mares up early in the evening, and turning them out
sometimes late in the morning. They *may* be right.
My advice is to adopt quite a different course, which I
have always found successful, viz. to give them the shed
or loose box during the heat of the day, with nice cool
soil, turning them out in the afternoon, and leaving
them at certain periods of the year out all night; but,
most certainly, never during the summer season shutting
them up before dusk, and always taking care that they
should be out at sunrise—for those are the very periods
at which they feed best, and at which their foals play about

and take their own exercise; moreover, the latter course makes the produce more hardy and healthy. The most desirable bedding during the day-time in summer is turf-mould; if it cannot be had, tan: the latter I am not so certain about, never having *tried* it. I merely fancy it should, or ought to be, a good substitute for the former, which is cooling, and otherwise beneficial to the feet. As proof in favour of the system recommended, any visitor will find that brood mares feed better at night than at any other time, for he will hear them literally mowing the grass like a scythe; whereas at mid-day he will find them under a hedge, covered with flies, teasing and biting them. One would fancy such symptoms would dictate the propriety of adopting the " shutting-up " system during the day, and *vice versâ*. I have seen mares at celebrated establishments, absolutely in mid-summer, locked up *for the night* at five o'clock in the afternoon.

The contrast in the ideas of parties as to prices and value of brood mares is not only extraordinary, but most amusing. Several years ago a gentleman in Ireland took it into his head to breed thoroughbred horses for sale, although he knew as much about the animal, in any shape or form, as Heliogabalus did about economy: his idea being, that he ought to get a *few good ones* for one hundred pounds, *as they had done racing*. " I see no reason," said he, " why I should not breed, and make it pay." One fine day certain persons, resolving upon having a "lark," caused a letter to be written by a friend in Liverpool, stating that he had little doubt he could procure for him the dam of the magnificent ' Canezou ' (then in the zenith of her glory), heavy with foal to 'Melbourne,' for fifty pounds. The would-be breeder came to consult his friends, and after mature deliberation replied,—"That if the mare were delivered in Dublin free of expense, and with foal, he should not mind giving thirty pounds for her." It is hardly necessary to mention the reception any proposal to purchase at any price would have met with at Knowsley, if such a subject had been broached. Our hero spent the intermediate anxious moments in informing his circle of acquaintances "that he was about to add this valuable animal to his stud of brood mares!" To my perfect knowledge, the person referred to would not afford his animal the opportunity of consuming a sack of oats quarterly, yet he would talk for hours of his expectations of being recorded amongst the most fashionable breeders of the day. As a matter of course, from that hour to the present he has never bred one that could win a saddle.

At the same time there are plenty of valuable mares to be had for moderate prices. Amongst others I purchased the following at the sums named:—'Dawn of Day' (dam of 'Twilight' and 'Rising Sun,' &c.), when a yearling, and perfectly sound, for 20l.; 'The Countess,' own sister to 'The Baron,' a splendid mare (dam of 'Lady Kingston'), four years old, and perfectly sound, 150l.; 'Thorn' (dam of 'Sprig of Shillelagh,' &c.), seven years old, and as sound as when foaled, with a yearling by 'Irish Birdcatcher,' and a colt foal at foot, 150l. for the lot; 'Queen Bee,' six years old, and sound (dam of 'Roman Bee,' &c.), with a colt at foot, for 40l.; 'Devotion' (dam of 'Mount Zion,' and 'The Druid'), 50l. All the foregoing as fresh and sound as the day they were foaled. A great deal depends upon the market one goes to, &c.

However people may differ upon other questions relating to breeding and the various strains of blood, it can hardly be denied that, at the present time, 'The 'Touchstone,' 'Melbourne,' 'Irish Birdcatcher,' and 'Sweetmeat' mares, have proved very superior, not only in point of numbers, but in quality; I have already remarked upon the successful results of the close alliances between the 'Whalebone' pair: yet still, although we find those animals winning frequently over short courses, and, no doubt, occasionally over long ones, it by no means follows that a very great improvement, especially in the latter respect, might not be made by crossing more frequently with some other strains of blood, such as 'Lancrcost,' 'Voltigeur,' and their sons. The principal object, however, of breeders at present appears to be (and a very natural one, too), to ensure a quick and almost certain return, rather than to incur risk by speculation.

If in-and-in breeding be so desirable, why not go the whole hog and test it? But then comes the question, Who will forfeit the "bird in the hand" to look for "two in the bush?" My reply is, Those who can afford the risk, and who in all other respects spare no expense: they are few, it must be confessed; yet there are a few. And this hint is merely offered for their consideration as an experiment, as well as that of adopting the course exactly opposite, and trying what a good "mixture" would accomplish; for admitting, if we go back a few generations, we find most of the animals of the present day nearly allied, through one strain or other, still there

arc many distinct and equally valuable and distinguished
within the past thirty years, whose lamp of life and
patronage has been for a long time flickering, and appears likely to be shortly extinguished, notwithstanding
its having in its day shone most brilliantly.    An instance of in-and-in breeding is found in 'Manœuvre'
(dam of 'Wallace' and 'Lioness'), being by 'Rector,'
son of 'Muley,' and her dam by 'Muley.'

It is the fashion to talk of crossing, with a view to
shapes, &c., that would suit blood, and so forth.    All
things, in the present "sensation" times, which used to
be termed "days of reform," go with the fashion of the
day.    Formerly, experienced judges would laugh at the
close alliances which we find now-a-days to prove most
successful; and really, taking into consideration the
"curious" things we see and hear of, day after day, we
must not be surprised if we find the old system I have
referred to revived.

In offering a few suggestions as to the brood mares
of the present day, both tried as racehorses and at stud,
and also upon those as yet untried in the latter respect, I merely do so, as far as the former are concerned,
with a view to point the reader's attention to cases of
failure, as well as to those where they have proved successful : as also to remind him, that the descendants of those
mares ought to be worthy the notice of breeders, whenever
an opportunity is offered to purchase.    With regard to the
untried division, I simply select them from memory of
their shapes and qualities; a few especial favourites of
mine having struck me, during their racing career, as
animals of superior merit, and most likely to prove successful at stud, not alone as to pedigree and performances,

out in every other respect: freely admitting that there may be many others equally good.

The tried mares require little comment from me, further than to refer the reader to their pedigree, and the "principal" produce, which will show what crosses of blood have proved successful. I shall venture, however, a few remarks generally upon "my chosen few," of the yet untried young mares of the present day; which, in my humble opinion, will be heard of to advantage at a future day, not far distant.

In making a selection of the picked, "tried," brood mares, it does not require much study or consideration to arrive at the conclusion that the following call for especial notice, many having distinguished themselves at stud, and many during their racing career; although it will be seen that some have not proved successful on the turf, yet did so as brood mares: a fact not at all to be wondered at. Amongst other reasons, because they have, in many instances, gone to the harem fresh and sound, and in a proper, vigorous, and healthy condition.

# SYNOPTICAL TABLE

OF

## THE TRIED BROOD MARES OF THE DAY,

WITH

THEIR PEDIGREE, BEST PRODUCE, AND THE SIRES OF SUCH PRODUCE.

| BROOD MARE. | PEDIGREE. | BEST PRODUCE. |
|---|---|---|
| ACHARANTHES .. | by Thirsk, dam Amaranth, by Bay Middleton .. | Gladiolus, by Kingston. |
| ADA MARY .. | by Bay Middleton, d. by Tramp .. .. | Adamas, by Touchstone. |
| AGNES .. | by Pantaloon, d. Black Agnes, by Velocipede .. | Queen of the Vale & Evelina, by King Tom. |
| ALEXINA .. | by Hetman Platoff, d. Y. Medora, by Prince .. | Mary, by Idleboy. |
| APHRODITÈ .. | by Bay Middleton, d. Venus, by Sir Hercules .. | Argonaut, by Stockwell. |
| APRICOT .. | by Sir Hercules, d. Preserve .. .. | Fravola, by Orlando. |
| ARROW, THE .. | by Slane, d. Southdown, by Defence .. | Cambuscan, by Newminster. |
| AS YOU LIKE IT .. | by Touchstone, d. Emma, by Whisker .. | Audrey, by Stockwell. |
| ATHENA PALLAS .. | by Irish Birdcatcher, d. Minerva, by Muley Moloch .. | Apollyon, by Mildew. / Neptunus, by Weatherbit. |
| ATHOL BROSE .. | by Orlando, d. Haggish, by Bay Middleton .. | The Hadji, by Faugh-a-Ballagh. |
| ATTACK .. | by Touchstone, d. Ghuznee, by Pantaloon .. | Bevis, by Buckthorn. |
| AVONMORE .. | by Old England, d. Haitoe, by Sir Hercules .. | Fitz-Avon, by Girikol. / Avondale, by Ratan. |
| BABETTE .. | by Faugh-a-Ballagh, d. Barbarina, by Plenipotentiary .. | Bathilde, by Stockwell. |

| | | |
|---|---|---|
| BARBARIAN, mare .. | her d. Allegrette, by St. Luke .. .. .. | Anfield, by Confessor. |
| BARBATA .. .. | by The Bard, d. Vitula, by Voltaire . .. .. | { Moorhen, by Chanticleer.<br>{ Lopcatcher, by Birdcatcher. |
| BARBELLE .. .. | by Sandbeck, d. Darioletta .. .. .. | { Flying Dutchman, by Bay Middleton<br>{ Van Tromp, by Lanercost.<br>{ Zuyder Zee, by Orlando. |
| BASSISHAW .. .. | by The Prime Warden, d. Miss Whinny, by Sir Hercules .. | { Ben Webster, by Barnton.<br>{ Isoline, by Ethelbert. |
| BAY MIDDLETON, mare | d. Vitula, by Voltaire .. .. .. | Orestes, by Orlando. |
| BAY CELIA .. .. | by Orlando, d. Hersey, by Glaucus .. .. | The Duke, by Stockwell. |
| BIRDCATCHER, mare.. | d. Nan Darrell, by Inheritor .. .. | Vedette, by Voltigeur. |
| BIRTHDAY .. .. | by Pantaloon, d. Honoria .. .. .. | Lupellus and Lupus, by Loup-Garou. |
| BLACK BESS .. .. | by Ratcatcher, d. Polydora, by Priam .. | Highwayman, by Cotherstone. |
| BLACKBIRD .. .. | by Irish Birdcatcher, d. Aspen, by Abbas Mirza .. | Ivanhoff, by Muscovite. |
| BLAME .. .. | by Touchstone, d. Maid of Lyme, by Tomboy .. | Chère Amie, by Sweetmeat. |
| BLISTER .. .. | by Bay Middleton, d. Hope, by Touchstone .. | Mainstone, by King Tom. |
| BLONDELLE .. .. | by Orlando, d. Sister to Lugwardine, by Bobadil .. | Biondella, by Flying Dutchman. |
| BLOOMING HEATHER | by Melbourne, d. Queen Mary, by Gladiator .. | Goise, by King Tom. |
| BLUE BELL .. .. | by Ion, d. Blanche of Devan, by Bedlamite .. | Eurydice, by Woolwich. |
| BRAVERY .. .. | by Gameboy, d. Ennui, by Bay Middleton .. | Rupee, by The Nabob. |
| BRIBERY .. .. | by The Libel, d. Split Vote, by St. Luke .. | St. Albans, by Stockwell. |
| BRIDAL .. .. | by Bay Middleton, d. Gold Pin .. .. | Special License, by Cossack. |
| BRIDLE .. .. | by the Saddler, d. Moneda, by Taurus .. .. | Habena, by Irish Birdcatcher. |
| BURLESQUE .. .. | by Touchstone, d. Maid of Honour, by Champion .. | Buckstone, by Voltigeur. |
| BUZZ .. .. | by Muley Moloch, d. Scandal .. .. | Amazonian, by Orlando. |
| CALCAVELLA .. | by Irish Birdcatcher, d. Caroline, by Drone .. | Money Spinner, by Marsyas. |
| CAMEL, mare .. | d. Lady Elizabeth .. .. | Muscovite, by Hetman Platoff. |

| BROOD MARE. | PEDIGREE. | BEST PRODUCE. |
| --- | --- | --- |
| CANEZOU .. .. | by Melbourne, d. Madame Pèlerine, by Velocipede .. | { Fazzoletto, by Orlando. Cape Flyaway, by Flying Dutchman. |
| CAPRICE .. .. | by Coronation, d. Lælia, by Sheet-Anchor .. .. | Union Jack, by Ivan. |
| CASTANETTE .. .. | by Don John, d. Nickname, by Ishmael .. .. | Fandango, by Barnton. |
| CATHERINE .. .. | by Don John, d. Arachne, by Filho da Puta .. .. | Butterfly, by Turnus. |
| CATHERINE HAYES .. | by Lanercost, d. Constance, by Partisan .. .. | { Trovatore, by Birdcatcher. Costa, by The Baron. |
| CHARITY .. .. | by Melbourne, d. Bencvolence, by Figaro .. .. | Limosina, by Newminster. |
| CHASEAWAY .. .. | by Harkaway, d. Victoria, by Philip the First .. | Bally Edmond, by Bantam. |
| CHEROKEE .. .. | by Redshank, d. by Middleton .. .. .. | North Lincoln, by Pylades. |
| CINIZELLI .. .. | by Touchstone, d. Brocade, by Pantaloon .. .. | { The Marquis, by Stockwell. Marchioness, by Melbourne. |
| CLEMENTINA .. .. | by Venison, d. Cobweb, by Phantom .. .. | Melissa, by Orlando. |
| COMFIT .. .. | by Sweetmeat, d. Troica, by Lanercost .. .. | Carbineer, by Rifleman. |
| COOK, THE .. .. | by Irish Birdcatcher, d. the Popping Piece, by Sir Hercules or Napoleon .. .. .. .. .. | { Croagh Patrick, by Mountain Deer. |
| COQUETTE .. .. | by Napier, d. Lælia, by Sheet-Anchor .. .. | Caroline, by Ivan. |
| CORDELIA .. .. | by Red Deer, d. Emilia, by Y. Emilius .. .. | Thunderbolt, by Stockwell. |
| COUNTESS OF BUR-LINGTON .. .. | } by Touchstone, d. Lady Emily, by Muley Moloch .. | Cavendish & Hartington, by Voltigeur. |
| CROSSLANES .. .. | by Slane, d. Diversion, by Defence.. .. .. | Alvediston, by Tadmor. |
| CURE, mare .. .. | d. Elphine, by Emilius .. .. | The Wizard, by West Australian. |
| CYMBA .. .. | by Melbourne, d. Skiff, by Sheet-Anchor .. .. | Barchettina, by Pelion. |
| DAISY .. .. | by Touchstone, d. Abaft, by Sheet-Anchor .. .. | Paris, by Mildew. |
| DAME COSSER .. .. | by Voltaire, d. by Whisker .. .. .. | Duneany, by Flying Dutchman. |
| DAUGHTER OF THE STAR | by Kremlin, d. Evening Star .. .. .. | Tomyris, by King Tom. |
| DEFAMATION.. .. | by Iago, d. Caricature, by Pantaloon .. .. | Saccharometer, by Sweetmeat. |

| Name | Sire / Dam | Produce |
|---|---|---|
| DEFENCELESS | by Defence, d. by Cain | Caractacus, by Kingston. |
| DEIOPEIA | by Defence, d. Romaike, by Rowton | Canace, by King Tom. |
| DIADEM | by Coronation, d. by Bay Middleton | Tiara, by Woolwich. |
| DINAH | by Clarion, d. Rebecca, by Sir Hercules | Commotion, by Alarm. |
| DOE, THE | by Melbourne, d. Actual, by Actæon | Borderer, by Joc o' Lot. |
| ELLEN MIDDLETON | by Bay Middleton, d. Myrrah, by Malek | Wild Dayrell, by Ion. |
| ELLERDALE | by Lanercost, d. by Tomboy | Ellington and Gildermire, by The Flying Dutchman. Summerside, by West Australian. Wardernarske, by Irish Birdcatcher. |
| ELSPETH | by Irish Birdcatcher, d. Blue Bonnet, by Touchstone | Drumour, by Big Jerry or Weatherbit. |
| EMERALD | by Defence, d. Emiliana, by Emilius | Mentmore, by Melbourne. King of Diamonds, by King Tom. |
| EMPRESS | by Emilius, d. Mangel-Wurzel | Autocrat, by Bay Middleton. |
| ENNUI | by Bay Middleton, d. Blue Devils, by Velocipede.. | Saunterer, by Birdcatcher. Loiterer, by Stockwell. |
| EQUATION | by Emilius, d. Maria, by Whisker.. | Exact, by Irish Birdcatcher. Diophantus, by Orlando. |
| ESPOIR | by Liverpool, d. Espérance, by Lapdog | Ethelbert, by Faugh-a-Ballagh. Brown Duchess, by The Flying Dutchman. |
| EULOGY | by Euclid, her d. Martha Lynn, by Mulatto | Impérieuse and Impératrice, by Orlando. |
| EXACT | by Irish Birdcatcher, d. Equation, by Emilius | Q. E. D., by Kingston. |
| FAIR HELEN | by Pantaloon, d. Rebecca, by Lottery | Lord of the Isles, by Touchstone. |
| FIRST RATE | by Melbourne, d. Ninny, by Bedlamite | Red Eagle, by Irish Birdcatcher. |
| FLASH OF LIGHTNING | by Velocipede, d. Dido | Electric and Battery, by Fallow Buck. |
| FLAX | by Surplice, d. Odessa, by Sultan | Queen Bertha, by Kingston. |

| BROOD MARE. | PEDIGREE. | BEST PRODUCE. |
| --- | --- | --- |
| Forget-Me-Not | by Hetman Platoff, d. Oblivion, by | Daniel O'Rourke, by Irish Birdcatcher. Vergiss-mein-Nicht, by Flying Dutchman. |
| Foinualla | by Irish Birdcatcher, d. Brandy Bet, by Canteen | Minepie, by Sweetmeat. Kingstown, by Tearaway. |
| Gaiety | by Touchstone, d. Caststeel, by Whisker | Ganester, by Cossack. |
| Gardham, mare | d. by Langar | Skirmisher, by Voltigeur. |
| Gipsy Queen | by Tomboy, d. Lady Moore Carew, by Trump | High Treason, by Mildew. |
| Gruyère | by Verulam, d. Jennala, by Touchstone | Parmesan, by Sweetmeat. |
| Hag | by Cowl, d. Cheshire Witch, by Pantaloon | Gallus, by Chanticleer. |
| Haricot | by Mango or Lanercost, d. Queen Mary, by Gladiator | Caller-Ou, by Stockwell. Cramond, by Andover. |
| Hawise | by Jereed, d. Sister to Hornsea, by Velocipede | Nancy, by Pompey. Erin-go-Bragh, by Sprig of Shillelagh. |
| Honey Dear | by Plenipotentiary, d. My Dear, by Bay Middleton | The Wild Huntsman, by Harkaway. Oxford, by Irish Birdcatcher. |
| Hybla | by The Provost, d. Otisina, by Liverpool | Mincemeat, by Sweetmeat. Kettledrum, by Rataplan. |
| Ignorance | by The Little Known, d. Bohémienne, by Confederate | Ignoramus, by Flying Dutchman. |
| Incurable | by The Cure, d. Elphine, by Emilius | Wingrave, by King Tom. |
| Iris | by Ithuriel, d. Miss Bowe, by Catton | Blondin, by Voltigeur. |
| Irish Queen | by Harkaway, d. Emily, by Pantaloon | Sweet Sauce and Sugar Plum, by Sweetmeat. Ace of Clubs, by Stockwell. |
| Jamaica | by Liverpool, d. Preserve | Yellow Jack, by Irish Birdcatcher. Cariboo, by Venison. |
| Jeremy Diddler, mare | d. Equation, by Emilius | Rogerthorpe, by The Hero. Sir Roger, by Saunterer. |

X

| BROOD MARE. | PEDIGREE. | BEST PRODUCE. |
| --- | --- | --- |
| MARGARET OF ANJOU .. | by Touchstone, d. Margaret, by Margrave.. | { Confectioner, by Sweetmeat. / Zambesi, by Saunterer. |
| MARION .. .. | by St. Martin, d. Rebecca, by Lottery .. | Marionette, by Touchstone. |
| MARMALADE .. | by Sweetmeat, d. Theano, by Waverley .. | Dundee, by Lord of the Isles. |
| MARY .. .. | by Melbourne, d. Jill, by Touchstone .. | Arrogante, by Stockwell. |
| MARY AISLABIE .. | by Malcolm, d. by Actæon .. .. | { Newcastle, by Newminster. / Lady Ripon, by Stockwell. |
| MARY COPE .. .. | by The Flying Dutchman, d. Blue Bonnet, by Touchstone | Marigold, by Teddington. |
| MEEANEE .. .. | by Touchstone, d. Ghuznee, by Pantaloon.. | Emily & Lady Augusta, by Stockwell. |
| MENTMORE LASS .. | by Melbourne, d. Emerald, by Defence .. | Evelina, by King Tom. |
| MERRY BIRD .. | by Irish Birdcatcher, d. Miss Castling, by Inheritor | Zetland, by Voltigeur. |
| MIDIA.. .. | by Scutari, d. Marinella, by Soothsayer .. | { Avalanche and Hurricane, by Wild Dayrell. |
| MISERRIMA .. | by Pantaloon, d. Phryne, by Touchstone .. | Mæstissima, by Pyrrhus the First. |
| MISS AGNES .. | by Irish Birdcatcher, d. Agnes, by Clarion .. | Little Agnes, by The Cure. |
| MISS ANN .. | by The Little Known, d. Bay Missy, by Bay Middleton .. | Scottish Chief, by Lord of the Isles. |
| MISS BATTY .. | by The Hydra, d. Sister to Malibran, by Muley .. | { Emblem and Emblematic, by Teddington. |
| MISS JULIA BENNETT | by Muley Moloch, d. Patty.. .. | Vandermulin, by Van Tromp. |
| MISS SARAH .. | by Don John, d. Miss Sarah, by Gladiator .. | Anonyma, by Stockwell. |
| MISS SELLON .. | by Cowl, d. Belle Dame, by Belshazzar .. | Seclusion, by Tadmor. |
| MISS TWICKENHAM .. | by Rockingham, d. Electress, by Election .. | Teddington, by Orlando. |
| NICOTINE .. | by Ion, d. Prussic Acid, by Voltaire .. | { Simple Simon, by Woodpigeon. / Lime-flower, by Knight of St. George. |
| NINA .. .. | by Cotherstone, d. Annette, by Priam .. | Walloon, by Flying Dutchman. |
| ORLANDO, mare .. | d. Brown Bess, by Camel .. | General Peel, by Y. Melbourne. |

| | | |
|---|---|---|
| PALMA .. .. | by Emilius, d. Francesca .. .. | Adventurer, by Newminster. |
| PALMISTRY .. | by Sleight-of-Hand, d. by Lottery .. | St. Giles, by Wonnersley. |
| PALMYRA .. .. | by Sultan, d. Hester, by Camel .. | Tadmor, by Ion. / Aleppo, by Alarm. |
| PANTALOON, mare .. | d. Daphne, by Laurel | Leamington, by Faugh-a-Ballagh. |
| PARADIGM .. .. | by Paragone, d. Ellen Horne, by Redshank | King-at-Arms, Man-at-Arms, and Blue Mantle, by Kington. |
| PASTRYCOOK .. .. | by Sweetmeat, d. Theano, by Waverley | Paste, by Kingston. |
| PAULINE .. .. | by The Emperor, d. Bettina, by Sultan .. | Arcadia, by Arthur Wellesley. |
| PEASANT GIRL .. | by The Major, d. Glance, by Waxy Pope .. | Joskin, by West Australian. / Lord Burleigh, by Prime Minister. |
| PEGGY .. .. | by Muley Moloch, d. Fanny, by Jerry .. | Musjid, by Newminster. |
| PETTICOAT .. .. | by Pantaloon, d. Camp-Follower, by The Colonel | Gunner, by Artillery. |
| PHEMY .. .. | by Touchstone, d. Phœbe, by Lamplighter .. | Russley, by Oulston. |
| PHŒBE .. .. | by Touchstone, d. Netherton Maid, by Sheet-Anchor | Big Ben, by Ethelbert. |
| PICCAROON, mare .. | (called The Hipped Mare) .. .. | Old Calabar, by King Tom. |
| PICCAROON, mare .. | d. Bonny Bonnet, by Muley Moloch | Tasmania, by Melbourne. |
| PICNIC .. .. | by Glaucus, d. Estelle .. .. | Mayonaise, by Teddington. |
| PLENIPOTENTIARY, mare | d. Myrrha (Arab's dam) .. | Bashi Bazouk, by Vortex. |
| POCAHONTAS .. .. | by Glencoe, d. Marpessa, by Muley | Stockwell and Rataplan, by The Baron-King Tom, by Harkaway. / Knight of Kars, by Nutwith. / Kt. of St. Patrick, by Kt. of St. George. / Ayacanora, by Irish Birdcatcher. / Automaton, by Ambrose. |
| POLYDORA .. .. | by Priam, d. Manto .. .. | Pandora, by Cotherstone. |
| PRAIRIE BIRD .. | by Gladiator, d. Valentine, by Voltaire | The Trapper, by Ion. |
| PRAIRIE BIRD .. | by Touchstone, d. Zillah, by Reveller | Bird on the Wing, by Irish Birdcatcher. |

| BROOD MARE. | PEDIGREE. | BEST PRODUCE. |
|---|---|---|
| PRIESTESS .. .. | by The Doctor, d. Biddy, by Bran .. .. | Dulcibella, by Voltigeur. / Romulus, by The Flying Dutchman. |
| PRINCESS .. .. | by Merry Monarch, d. Queen Charlotte, by Elis .. .. | Nutbourne, by Nabob. / Bertha, by Stockwell. |
| PYRRHUS THE FIRST, mare .. .. | d. Miss Whip, by the Provost .. .. | The Drake, by Stockwell. |
| QUEEN, THE .. .. | by Iago, d. Duchess of Kent, by Belshazzar .. | Dictator, by The Cure. |
| QUEEN BEE .. .. | by Harkaway, d. Calcavella, by Irish Birdcatcher .. | Roman Bee, by Birdcatcher or Artillery |
| QUEEN BEE .. .. | by Amorino, d. Mayfly, by Emilius .. | Cerintha, by Newminster. |
| QUEEN MARY .. .. | by Gladiator, d. by Plenipotentiary.. .. .. | Haricot, by Mango or Lanercost. / Blink Bonny, by Melbourne. |
|  | d. by Melbourne .. .. .. | Solomon, by Teddington. |
| RATAN, mare .. .. | by Annandale, d. Messalina, by Bay Middleton .. | Stampedo, by Alarm. |
| REPENTANCE.. .. | by Bay Middleton, d. Moss Rose, by Blacklock .. | Wild Rose, by Surplice. |
| ROSE OF CASHMERE .. | by Ithuriel, d. by Partisan .. .. .. | Horror, by Wild Dayrell. |
| SALLY .. .. .. | by Gameboy, d. by Muley Moloch .. .. | Indifference, by Irish Birdcatcher. |
| SANGFROID .. .. | by Touchstone, d. Ghuznee, by Pantaloon .. .. | King of Kent, by Ethelbert. |
| SCALADE .. .. | d. Miss Eliza, by Humphrey Clinker .. .. | Victor, by Vindex. |
| SCROGGINS, mare | by Slane, d. Seabreeze, by Paulowitz .. .. | Laodamia, by Pyrrhus the First. |
| SEAL .. .. .. | by Orlando, d. Ladye of Silverkeld Well, by Velocipede.. | Caterer, by Stockwell. |
| SELINA .. .. | by Young Priam, d. Miss Bucktrout, by Perion or Tomboy | Kildonan, by Newminster. |
| SHAMROCK .. .. | by Birdcatcher, d. Wasp, by Muley Moloch .. .. | Donna del Lago, by Lord of the Isles. |
| SHOT .. .. .. | her d. Electress, by Election .. .. | British Remedy, and Gin, by Orlando. |
| SIR HERCULES, mare.. | by Orlando, d. Princess Alice, by Liverpool .. .. | Tom Fool, by King Tom. |
| SKIT .. .. .. | by Melbourne, d. Volley, by Voltigeur .. .. | Lord Clifden, by Newminster. / Lady Clifden, by Surplice. |
| SLAVE, THE .. .. | d. Hamptonia, by Hampton.. .. | Elcho, by Rifleman. |
| SLEIGHT-OF-HAND, mare |  |  |

| | | |
|---|---|---|
| SLEIGHT-OF-HAND, mare | d. (1843) by Recovery | King of Hearts, by Daniel O'Rourke |
| SNOWDROP | by Heron, d. Fairy, by Filho da Puta | Gemma di Vergy, by Sir Hercules. |
| SORTIE | by Melbourne, d. Escalade, by Touchstone | Stockade, by Stockwell. |
| SPRIGHTLY | by Gladiator, d. Sprite, by The Mole | Physician, by Chanticleer. |
| STAMP | by Emilius, d. Receipt | Fitz-Roland, by Orlando. Queen's Head, by Bay Middleton. Exchequer, by Stockwell. |
| STEEL PEN | by Irish Birdcatcher, d. Needle, by Lanercost | Magnum Bonum, by Lanercost. |
| STOLEN MOMENTS | by Melbourne, d. Lady Elizabeth, by Sleight-of-Hand | Lady Trespass, by Irish Birdcatcher. Secret Treasure, by Daniel O'Rourke. |
| SULTANA | by Hetman Platoff, d. Green Mantle, by Sultan | Theodora, by Orlando. |
| SUNFLOWER | by Bay Middleton, d. Io, by Taurus | Sunbean, Rainbow, Northern Light, by Chanticleer. |
| SWEET PEA | by Touchstone, d. Pink Bonnet, by Lanercost | Madame Cliquot (afterwards called Forget-Me-Not), by Burgundy. |
| SYBIL | by The Ugly Buck, d. Sylph, by Filho da Puta | Oldminster, by Newminster. Tim Whiffler, by Van Galen. |
| SYLPHINE | by Touchstone, d. Mountain Sylph, by Belshazzar | Camarino, by West Australian. |
| TEETOTUM | by Touchstone, d. Versatility | Asteroid, by Stockwell. |
| TERMAGANT | by Cotherstone, d. Virago, by Velocipede | Sacerdos, by Surplice. |
| THEMIS | by Touchstone, d. Rectitude | Sedbury, by The Cure or Sweetmeat. |
| THORN | by Harkaway, d. Whiterose, by Plenipo | Sprig of Shillelagh, by Simoom. Shillelagh, by Teddington. |
| TITÆA | by Ion, d. Titania, by Emilius | Sawcutter, by Idleboy. |
| TOMYRIS | by Sesostris, d. by Glaucus | Eastern Princess, by Surplice. |
| TORMENT | by Alarm, d. by Glencoe | Adrasta, by Orlando. Tisiphone, by ditto. Tricolor and Touch-Me-Not, by Touchstone. |
| TRIANGLE | by Epirus, d. Fortress, by Defence | Marseillaise, by Bay Middleton. |

| BROOD MARE. | PEDIGREE. | BEST PRODUCE. |
| --- | --- | --- |
| TRICKSTRESS | by Sleight-of-Hand, d. Duchess of Kent, by Belshazzar | The Knave, by Orlando. |
| TROCHEE | by Venison, d. Iliona, by Priam | Romsey, by The Flying Dutchman. |
| TROUSSEAU | by Gameboy, d. Bridal, by Bay Middleton | Little Drummer and Peignoir, by Rataplan. |
| TRUTH | by The Libel, d. Miss Kitty Cockle, by Cudland | Sir Hercules, by Sir Hercules. |
| TWITTER | by Alarm, d. Little Finch, by Hornsea | Chirp, by Orlando. |
| URANIA | by Idleboy (son of Satan), d. Venus, by Langar | Amsterdam, by The Flying Dutchman. |
| URSULINE | by Surplice, d. Treacherous, by Harkaway | Coup d'Etat, by Leamington. |
| UTOPIA | by Jerry, d. Turquoise | King of Utopia, by King of Trumps. |
| VALENTINIA | by Velocipede, d. Jane, by Moses | Clotilde, by Touchstone. / Cecilia, by Windhound. |
| VARSOVIANA | by Ion, d. by Langar | Nemesis, by Newminster. |
| VEST | by Cotherstone, d. Cloak, by Rockingham | Investment, by Wild Dayrell. |
| VÉSUVIENNE | by Gladiator, d. Venus, by Sir Hercules | Crater and Lava, by Orlando. |
| VEXATION | by Touchstone, d. Vat, by Langar | Defiance, by Old England. |
| VIVANDIÈRE | by Voltaire, d. Martha Lynn, by Mulatto | Cantine, Provision, and La Fille du Régiment, by Orlando. |
| VOLATILE | by Buckthorn, d. Jocose, by Pantaloon | Carnival, by Sweetmeat. |
| VOLLY | by Voltaire, d. Martha Lynn, by Mulatto | The Slave, by Melbourne. / Little Lady, by Orlando. |
| WIASMA | by Hetman Platoff, d. Mickleton Maid, by Velocipede | Mouravieff, by Pyrrhus the First. / Viatka, by Teddington. / Klarikoff, by De Clarc. |
| WICKET | by Stumps, d. by Phantom | Rattlebone, by Cruiser. |
| WOLDSMAID | by Hampton, d. Sister to Grey Momus | Moorcock, by Chanticleer. |
| YARD-ARM | by Sheet-Anchor, d. Fanny Kemble, by Paulowitz | Gunboat and Lifeboat, by Sir Hercules. |
| ZENOBIA | by Slane, d. Palmyra, by Sultan | Aurelian, by Stockwell. |

The brood mares now at stud, according to official returns, number about 2250. The foregoing table represents some of the tried mares, which number about 100; many of them having merely produced animals of moderate form; and very few having distinguished themselves as dams of first-class animals. If we except about twenty of the tried division, the others will be found either to have been perfect failures, or with but one recommendation; having produced what may be termed "chance produce." And many of those at stud, that do not rank according to the rules of breeding and racing, as tried brood mares, are and have, in point of fact, proved themselves far superior; for although their stock may not be qualified as winners of even a fifty-pound plate, yet they have been "seconds," and otherwise proved themselves superior animals in far better company. If we select—

| | |
|---|---|
| Pocahontas | Midia |
| Alice Hawthorne | Equation |
| Ellerdale | Espoir |
| Barbelle | Palmyra |
| Cinizelli | Queen Mary |
| Mountain Sylph | The Arrow |
| Hybla | Wiasma |
| Eulogy | Canezou, |

we have about the only ones that can be looked upon as really successful during their career at stud, or as entitled to be viewed in the light of "really good brood mares;" for they have been successful in producing not only several winners, but have done so from various sires and different crosses.

There are about one hundred 'Touchstone' mares at stud—double the number of any other, which must partly

account for their success. Others appear to have produced one first-class racehorse each; the rest being very
inferior, although crossed with various strains, and the
best blood and most fashionable sires of the day—namely,
'Ellen Middleton,' 'Bribery,' 'Burlesque,' 'Wizard's'
dam, 'Defenceless,' 'Fair Helen,' 'Hawise,' 'Irish
Queen,' 'Jocose,' 'Mainbrace,' 'Miss Twickenham,'
'Peggy.'

Then, again, there are several that have been perfect
"stars" during their racing career, and have so far proved
perfect failures at stud. Perhaps a more extraordinary
instance of the lottery of breeding cannot be furnished,
than 'Virago.' She has been put to stud since 1856.
Since which period she has been put to 'Orlando,' 'The
Flying Dutchman,' 'Stockwell,' 'Kingston,' &c. (a fair
chance, no doubt); and yet we have not heard of any
successful result.

Another extraordinary instance of failure is the 'Maid
of Masham;' one of the best, and certainly as sweet a mare
"as ever looked through a bridle." (I just missed being
her owner at five hundred guineas, "by one post,"
having written to accept her then owner's terms; but
she had gone the day previous.) She has been at stud
since 1852; and although put to 'Birdcatcher,' 'Orlando,'
'Teddington,' 'West Australian,' 'Young Melbourne,' &c.
she has not proved successful so far. And if ever there
was one instance, above any other, that could make an
impression (at least on my mind) of the "lottery,"
this is the one: for a more perfect specimen of a brood
mare in every shape, form, and respect, I never beheld,
and expect and hope yet to hear of her proving more successful at stud.

There are innumerable other instances where such splendid mares have failed at stud; and, curiously enough, the Oaks winners have proved most unsuccessful: for, commencing with the renowned 'Queen of Trumps,' with very few exceptions, indeed, they have been perfect failures.

Some mares appear to have proved more successful where a change from one sire to another has been made: as an example, 'Barbelle' produced 'The Flying Dutchman' by 'Bay Middleton;' 'Van Tromp,' by 'Lancercost;' 'Zuyder Zee,' by 'Orlando:' yet 'Vanderdecken,' own brother to 'The Dutchman,' was a very sorry sample, indeed, in every respect. 'Blue Bonnet' (winner of the Doncaster St. Leger, and by 'Touchstone') furnishes another instance of comparative failure at stud; for, although a first-class racehorse, notwithstanding every chance she only produced one really good animal, 'Mary Copp,' by 'The Flying Dutchman;' her other produce (some by the same sire) being comparatively moderate: and yet it is more than probable her sons and daughters will revive the good qualities of the family, more especially 'Mary Copp,' whose wonderful speed and other good qualities are most likely to render her a valuable addition to the stud; her daughter, 'Marigold,' having already afforded proof of her dam's good qualities (although got by that game racehorse, yet moderate sire, 'Teddington'). And if she does not further "give a good account of herself," I shall be much surprised, as she will, in my opinion, turn out one of the very best brood mares of the present day. Her *forte* was great speed; and if crossed judiciously, she can hardly fail to furnish some worthy scion of her distinguished race. She is one of my especial favourites. Her Goodwood running cannot easily be forgotten.

Suppose those extraordinary mares that have so distinguished themselves as racehorses had *never been trained,* but turned out at three or four years old to stud, what would have been the result? Can it be argued that their racing career tended to improve their valuable qualities as brood mares? The fact is, many superior ones are passed over, simply because they have not won an Oaks, a thousand-guineas stakes, &c., and are doomed to draw a Whitechapel cart, or condemned to some other uses vile. Some of the plainest-looking, moderate-sized mares, produce not only the best stock but the truest and best-shaped; for although they may not themselves exceed 15 hands high, or a little over, still, if put to stallions of general size and substance, &c., their produce is as large as others, and seldom, if ever, of that great, top-heavy, unwieldy form, coming so frequently from great, large mares. However good they may have proved during their racing career, the produce of the latter are seldom stayers, but come rolling in from one side of a course to the other, like a ship in a high sea, as tired as their pilot, without ballast, or the level, well-proportioned shapes of the produce of the moderate-sized, lengthy mare, "long and low."

In cases where mares produce different stock by the same sire, one perhaps a first-class racehorse, the rest bad, though all have been equally attended to and received proper care,—such, no doubt, are the instances which best supply proof of the uncertainty of breeding. Still, frequently the lottery is increased through neglect, mismanagement, and various causes; sometimes through "freaks of nature," to which all other animals are likewise subject. One might fairly ask, if 'Pyrrhus the First' could get such a mare as 'Virago,' 'Bay Middleton' such animals as the 'Flying Dutchman' and 'Andover;' or as to brood mares, if 'Beeswing' should have produced such an animal as 'Newminster,' 'Alice Hawthorne' 'Thormanby,' and 'Oulston,' why should they have otherwise proved so unsuccessful? Again, how is it that other sires and brood mares have been crossed with almost any other strain, and have proved so successful? There must be some reason. It is natural to assume that in the same hands the produce have been equally cared for. The very fact that in many instances some of these animals, when first put to stud, appear not to breed so well, yet improve with time, would in itself partly go to prove that severe and long training is the cause; but in any case there cannot be a question that there are certain "running strains," and that in those few instances where there is one "first class," the rest bad, or very indifferent, such animals are what may be termed "chance horses."

Without wishing for one moment to offer a remark in any way tending through prejudice to deteriorate the merits of sires, I cannot help expressing a very great dislike to 'Pyrrhus the First,' as a stallion, and believing a few more like him would be no acquisition. The late Sir Tatton Sykes, one would fancy, "assisted" towards advancing his *prestige*, having had a great number of his daughters in the harem, which did not prove more successful than his twenty-one Hampton and fifty-two Sleight-of-Hand mares—no doubt, as to numbers, sufficient in themselves to have laid the foundation of brilliant future success; still, consuming as much as an equal number of 'Touchstones,' 'Birdcatchers,' 'Sweetmeats,' and doubtless affording the owner quite as much amusement — it is to be regretted they did not likewise afford so thorough and popular a sportsman an opportunity of escorting their produce (as he did his namesake) into the Doncaster enclosures, which would have been hailed with an ovation sufficient to cause York Minster bells to ring: for, great as would have been the demonstrations in honour of the victory by a certain Scotch nobleman, the Baronet would have given him weight in Yorkshire. When parties commence breeding there is nothing like getting into a good strain, for it is extraordinary how they are carried on year after year (if they have got into a bad breed), rather than make a bold attempt to open the doors and get rid of the lot at any sacrifice, the first loss being at all times the best; yet they require moral courage, not frequently displayed, especially in regard to horseflesh.

Having referred to the tried mares, I now beg to place before the reader a few of my special favourites, as yet untried at stud; my opinion of their respective qualities being formed not on any particular grounds as to performances alone, but taking into consideration the combination of shapes, blood, and all the qualifications which may render such animals valuable and desirable to the breeder. They are as follows: —

| NAME.† | SIRE. | DAM. | GRANDSIRE. |
|---|---|---|---|
| *Ayacanora | Irish Birdcatcher | Pocahontas | Glencoe |
| Brown Duchess | The Flying Dutchman | Espoir | Liverpool |
| Cantatrice | Irish Birdcatcher | Catherine Hayes | Lanercost |
| Citron | Sweetmeat | Echidna | Economist |
| The Deformed | Burgundy or Harkaway | Welfare | Priam |
| Emily | Stockwell | Meeanee | Touchstone |
| Fravola | Orlando | Apricot | Sir Hercules |
| Governess | Chatham | ,, | Laurel |
| Habena | Irish Birdcatcher | Bridle | The Saddler |
| Katherine Logie | Flying Dutchman | Phryne | Touchstone |
| Lady Hawthorne | Windhound | Alice Hawthorne | Muley Moloch |
| Lady Palmerston | Melbourne | ,, | Pantaloon |
| *Marchioness d'Eu | Magpie | Echidna | Economist |
| Mincemeat | Sweetmeat | Hybla | The Provost |
| Nemesis | The Nob | The Arrow | Slave |
| Overreach | Irish Birdcatcher | Virginia | Rowton |
| Pandora | Cotherstone | Polydora | Priam |
| Peri | Irish Birdcatcher | Perdita | Langar |
| Queen of the Vale | King Tom | Agnes | Pantaloon |
| Rambling Katie | Melbourne | Phryne | Touchstone |
| *Rosabel | Newminster | ,, | Jereed |
| Rosa Bonheur | Touchstone | Boarding-school Miss | Plenipotentiary |
| Rupee | Nabob | Bravery | Gameboy |
| Stockade | Stockwell | Sortie | Melbourne |
| Summerside | West Australian | Ellerdale | Lanercost |
| Sunbeam | Chanticleer | Sunflower | Bay Middleton |
| Sweet Hawthorne | Sweetmeat | Alice Hawthorne | Muley Moloch |
| Terrific | Touchstone | Ghuznee | Pantaloon |
| Terrona | Touchstone | Alice Hawthorne | Muley Moloch |
| *Theodora | Orlando | Sultana | Hetman Platoff |
| Tunstall Maid | Touchstone | Tomboy Mare | Tomboy |
| *Typee | Touchstone | Boarding-school Miss | Plenipotentiary |
| Uzella | Sweetmeat | Welfare | Priam |

With regard to the blood of brood mares of the present day, and the crosses which suit (without entering into any detail of crosses particularly), I would recommend the

† Those mares marked with an asterisk (*) were in manuscript and selected before their produce won.

reader to adhere as far as possible to the following strains and mixtures, of course taking into account the running family in other respects; that is to say, as to other strains of blood a few generations back : first, to breed from 'Touchstone,' 'Melbourne,' 'Irish Birdcatcher,' 'Sweet-meat,' and 'King Tom' mares, in preference to others, as a general rule. No doubt the 'Bay Middleton' strain stands justly in the front rank, and has proved successful, con- . sidering the chances, in proportion.* The daughters of 'King Tom,' if fashionably bred on their dam's side, will, in my opinion, turn out most successful, and should be held in high esteem, not only from the value of 'The Whale-bone' blood through their splendid sire, but because in their veins flows the blood of the queen of brood mares, 'Pocahontas ;'† and I venture to predict that the 'King Tom' mares will prove second to none. Were I to suggest a cross, it would be with the 'Sweetmeat' strain, of the value of which 'Sweetsauce' has furnished so plain and striking a proof: his dam, 'The Irish Queen,' being by the same sire as 'The King,' viz. 'Harkaway.' 'Dundee'

---

* It is a curious fact that the 'Sweetmeat' and 'Melbourne' cross has been almost left entirely untried, although in the very few instances where an approach to it has been made, they have proved successful. The cross ought, in one respect at least, to be judicious ; viz. the symmetry of the one, with the fine bone and frame of the other, ought to prove successful. There is no strain can excel 'Melbourne' as fine samples of slashing racehorses — length, bone, &c ; and the blood is perhaps the best of the whole lot, in a general point of view.

† She has done more towards improvement in the racehorse than any mare ever foaled, and with 'Guiccioli,' dam of 'Irish Birdcatcher,' and 'Faugh-a-Ballagh,' has proved a mine of wealth to the turf.

also comes from the same cross, being a mixture of 'Whalebone' and 'Sweetmeat' strains.

No matter how the breeder may cross, or what blood he may select to build his hopes upon, to my mind there is one above all others that will improve "the mixture," whether it be for speed or stoutness, fine, slashing, and racing shape, size and action; and that strain is 'Pantaloon.'

Once more, in concluding my remarks upon brood mares, &c., let me impress upon the reader the absolute necessity of avoiding the "penny-wise-and-pound-foolish" principle, and of adhering to the *running family*, instead of one or two solitary exceptions, and a bunch of rubbish; and when he has got possession of them, to give them every chance—*the very best feeding, care, and attention.*

# THE STUD-FARM: ITS REGULATIONS AND REQUISITES.

There is not, perhaps, in any speculation, an element more precarious or expensive than that which forms the subject of the following remarks, which are made with a perfect consciousness that there are many persons more experienced in such matters; but still there are, no doubt, some who have not heretofore given their minds or attention to the subject, who might derive a few useful hints therefrom. The first step which should be taken is to find *good land, thoroughly drained,* naturally of a dry and fertile soil; the herbage mixed with clover, and the other various seeds to which the horse is partial: for if the reader will look at the coats (even in the middle of summer) of mares fed upon a wet or insufficiently drained pasture (the colour of which is invariably a very dark green, as well as coarse and sour), they will find them cold-looking and staring; and that, however abundant the crop may be, it does not possess that sweetness which renders it palatable to the animal,— a fact which becomes plainly visible and proved when we find certain portions of the pasture in long tufts untouched, and other spots eaten to the roots: therefore, the very first object of the breeder should be to select the best soil. Regard should be paid to comfort and necessaries in the shape of stabling, sheds, &c. ; and, independently of the usual houses, it is desirable—if the tract of land extends to any considerable distance, where the regular stables cannot at all hours be made use of—to have sheds erected at convenient spots, here and there, under which, during excessive heat in the day-time, according as the animals like, they can take shelter from the annoyance of flies—during which period they seldom feed—or from cold winds, &c.

The formation or structure of stabling for breeding purposes is a subject with which many are acquainted, and in the present day requires little comment, having been brought to perfection ; as will appear manifest to any visitor to the stud-paddocks at Knowsley, Eaton, and many other places. However, for the benefit of those who may not have had an opportunity of visiting these establishments, it may be useful to offer a few remarks.

The stables should be roomy, as large and lofty as possible, well ventilated, the doors (which, during the animals' absence, should be kept as much open as possible) being not only broad, but equally high in proportion, avoiding the possibility of accident to the animals when rushing suddenly in or out, through either striking their heads against the top, or otherwise injuring themselves by coming in contact with the sides, to prevent which wooden or leather-bound rollers should be placed at each. Extra ventilators should be placed *low down* at each side of the door, in a slanting manner, about half-a-foot in width and one foot from the ground, to be made use of when necessary. The flooring should be formed in such a manner that the bedding would remain as dry and clean as possible. In constructing the roof,

Y

particular care should be taken to have a sufficient space
between the slates and the ceiling, so as to prevent the
excessive heat of the sun on the former causing heat, or
the contrary effect during the cold in winter from frost
or snow, as each have great influence thereupon.  The
walls should be occasionally whitewashed, to banish any
nauseous smells or remains of distemper which might
exist; the mangers should be repeatedly cleansed from
the remains of the various descriptions of food previously
placed therein, and which become sour and disagreeable.
The wood-work should be lined with tin or zinc, in order
to prevent the animals biting, or learning to crib, or becoming wind-suckers, to which they are much inclined.

That the system of placing together a number of
mares at any period of the year (and which, I have
remarked, is frequently practised) is highly injurious,
becomes apparent in many ways: therefore my idea is,
that a reasonable tract of land allotted to a couple of
mares, with their separate houses, is most desirable : in
fact, the more horses and the more paddocks the better.
Especial care should be taken, as far as possible, to prevent
their connexion with strange animals, and that they
should be as distant from thoroughfares as possible ;
because every disturbance is to a certain degree injurious,
especially at periods when mares are either about to foal,
after foaling, or during the time they are going through
their trials.

Paddocks, if possible, should be formed in such a
manner that many animals could not have access to, or
their attention drawn towards others, especially when
strangers are likely to appear; therefore, walls or banks
are preferable to transparent palings or rails.  When the

latter form the partitions, the consequence is continual neighing, galloping, and annoyance to the whole establishment on the introduction of a new-comer: mares on the eve of foaling, yearlings full of flesh, galloping until they break their wind, spring curbs, or otherwise injure themselves; sires neighing and roaring,—in short, the whole establishment becomes a perfect bedlam.

Then, again, sires never should be kept near mares or young stock, but in some strictly private place, especially during the season, which is the most critical period for foaling mares, as well as for others: the continual noise attending the trials, and strangers perpetually coming and going, interfere most injuriously with the animals, preventing them from feeding, and in many instances from proving with foal, or probably causing some to pick or cast foal. One of the great objections to bundling a number of mares together, even in large tracts of pasture, is this: if they are on the eve of foaling, they frequently interfere with and annoy each other, rushing suddenly to bite, &c., for it is wonderful how they have their particular friends amongst their number, and their enemies also, never losing an opportunity on their approach of showing their dislike, and in the most cunning and vicious manner in many instances; for they keep on grazing until they get near enough, rushing at, biting, and kicking at the object of their dislike, causing the other to hastily jump round to avoid punishment, and perhaps receive serious injury, the foals also getting sometimes hurt.

On the other hand, in the case of barren mares recently sent to stud, perhaps twenty or thirty are in one field, and others are continually going and coming, which, in addition to the process of bringing each back and fro

during their trials, results in frequent disturbance to
the others, at a time when they should be kept as quiet
as possible; each in turn, during a certain period, viz.
before being done with their trials, going the round
perhaps of the whole lot, teasing them, sometimes getting
kicked for their trouble; from others finding a reciprocity
of feeling, which proves equally injurious to both, under
the circumstances. The question as to the size or shape
of the boxes is quite a secondary consideration, provided
they are moderately comfortable and properly ventilated,
compared to that of the quality of the soil, for there and
in feeding generally lies the foundation for bringing the
racehorse to perfection.

Perhaps the most important of all requisites in the
stud-farm is good water and plenty of it, running streams,
of course, being most desirable; but where they are not
found, large tubs should be placed in each field and kept
continually supplied with soft water. A water-barrel on
wheels is most useful for the purpose of filling such
tubs. However necessary regular feeding-hours may be,
such regularity becomes far more so as regards drinking
(there is nothing so frequent as inattention on this particular point); for brood mares especially, and more particularly in the hot weather, are fond of wetting their
mouths, and playing with the water.

Mares fed upon good soil seldom fail to show it, if
they are sound in constitution, free from worms, or in-
ternal disease. When the condition is perfect, their coats
in summer not only shine like satin, but bear a golden
tinge, resembling that in the peacock's feathers, although
I must confess I have seldom seen brood mares' condition brought to such perfection ; which is owing,

principally, to their own extraordinary health and sound constitution.

Parties desirous of improving or rendering a tract of pasture more convenient, which is of a square form, and has not been previously sufficiently subdivided, could not do better than erect a square building in the centre thereof, dividing it into four equally proportioned boxes, with a door opening into each fourth part of the field; the fences or divisions of which should not be transparent, as before mentioned, but should be formed of banks made from the earth at each side, with about three feet of the foundation breasted with common stone or bricks, gradually tapering towards the top, which should be at least eighteen inches wide, and sufficiently high to prevent the animals seeing or interfering with each other. It is, of course, more desirable to have brick or stone walls.

Corn-bruisers and boilers are likewise most necessary in a stud-farm; bruised oats being desirable for all animals, and boiled barley, linseed, turnips, bran, &c., being frequently required, especially during certain periods of the year, when mares are about to foal, and during winter and spring.

In a properly-conducted establishment there cannot be too much neatness or regularity as regards all the necessaries. A room for head-collars, &c., should be kept regularly fitted out. As to head-collars great care should be taken that they *fit properly*, neither too large nor too small; for it has frequently happened that, when horses scratch their heads with their hind legs, or otherwise, they become entangled and receive serious injury, in some instances breaking their necks, the head-collars being larger than necessary. Care should also be taken

not to leave in the stable buckets having iron hoops
or handles; for horses are inclined to paw at them, and
frequently injure themselves, the foot and leg becoming
fast between the iron hoop and the edges of the bucket:
the latter falling, the horse becomes frightened, springs
back, and by a sudden downward motion presses the iron
against the skin.  In the construction of the manger it is
desirable to have a sort of bar at each end, about five
or six inches from the extremities, to prevent the horse
pushing the oats out, which he is most likely to do, particularly when first placed before him, appearing to prefer
eating them from the ground, but generally wasting a
large quantity.

Inasmuch as the period from which horses take their
ages renders it desirable to have every precaution taken
to be as forward as possible with nourishment, and
the season of the year not usually supplying the quantity which might be wished for, care should be taken
to have a field with *a winter crop of rye-grass*, which
is much earlier than other pasture, and can be either
mown by the scythe or otherwise supplied, and which is
far more nutritious to the mare at foaling-time than
dried hay: the great drawback to the early foal being the
want of a sufficient quantity of milk, especially where the
dam may not be a good suckler.  Many experienced
persons maintain that an April foal is quite as good as
one dropped very much earlier, assigning as a reason the
want of a sufficient supply of milk or nourishment from the
dam, believing that the foal becomes retarded in its natural
growth : in short, all breeders should take care to have a
field of *early grass*, which is not more expensive than any
other, although far more desirable and profitable in every

respect. In the paddock the addition of a contrivance (which admits of various forms) is most useful for the purpose of enabling the foal to take his bruised oats, occasionally with a little moistened bran, or cut carrots without the interference of the dam (who, no matter how good a nurse, is certain to devour the entire before the foal has time to partake of any). The most simple plan is to place the food in the centre of a fenced-in inclosure, where the foals can have access to it without such interference, forming at a distance of about four feet a paling or obstruction to the mare; taking care at the same time that it should be well rounded, of proportionate height, so as not to injure the foal in passing under backwards and forwards.

Bearing in mind the number of contagious diseases with which horses are so frequently attacked, it is necessary to have, as far removed as possible from the general stabling, a sort of hospital, with well-ventilated and airy boxes, to which the horses upon the first appearance of sickness should be immediately removed, and the boxes from which they have been so changed should be forthwith whitewashed with lime and water, and otherwise cleansed and ventilated.

The stud-groom should be provided with the usual medicines and instruments, which are so frequently and so suddenly required, and might not on an emergency be easily procured; the want of which might be attended with very serious consequences to the owner. Any stud-groom possessing reasonable intelligence and ordinary experience will soon understand how to treat those cases which are so common in horse-flesh, and about which so much ridiculous fuss is frequently made.

Having made these few general remarks upon the absolute requisites to a properly-conducted breeding establishment — remarks which, no doubt, are superfluous to the numerous experienced parties who breed extensively, but which may prove useful to those who are about to commence, I would wish to impress upon the beginner, above all things, to be provided with those artificial or forced crops, *rye-grass* and *turnips*, which are likewise most desirable for brood mares at that season when they most require such nourishment: the latter strewn over the field are capital for producing milk; the former, being cut and placed before the mares, prevents its being trampled upon and wasted, at a period when it is so scarce and valuable. Boiled turnips mixed with bran, linseed, bruised oats, &c., by way of change at night, will be found most beneficial substitutes for the grass at that period of the year.

Where outlay is no object, perfection cannot be too dearly paid for in obtaining all the requisites referred to, and I would strongly recommend a beginner to remember, that to be "penny wise and pound foolish" is bad economy, especially in horsebreeding, which from the competition in the present day, as well as the remunerative prices realised by breeders, presents an open field for prosperous speculation.

There is, indeed, no speculation to which these remarks are so applicable as to that of horse-breeding ; it is one which peculiarly requires liberal ideas, and an entire absence of all petty considerations.

# CONCLUSION.

I FEEL it is now time to bring my observations on Turf Topics to a conclusion. In offering them to the public I am not influenced by any desire to acquire literary notoriety, or to be regarded as a successful author: my object is one far less egotistical, and I can with truth say, much more philanthropic. It is simply to extend to those whose youth and want of knowledge in racing matters may render the information valuable, the results of a lengthened and costly experience—to warn them against the dangers which beset their path—and yet, at the same time, to encourage them (if their circumstances permit of it) in the pursuit of that pastime which, of all others, is most fitted to add power to the mind and vigour to the body.

Even at the risk of being condemned as one prone to indulge in self-praise, I cannot refrain from adverting to the singular fulfilment of some of my predictions which has so recently taken place.

I allude, amongst others, to the success of 'Gladiateur' in the 2000-guineas race at Newmarket. If the reader will refer to the 142nd page of this Treatise, he will feel satisfied that the theories I there advanced have

z

been now proved to have been sound in their conception; and that subsequent events have fully justified opinions which were the offspring of a judgment that has been matured, as well by disappointment as by experience.

It may be that I have in some degree written so as to dishearten the man who truly loves the Turf; who is sufficiently favoured by fortune to indulge in its pleasures; but still who, from his peculiar idiosyncracy, may not be able to bear up against reverses. If I have done so, I did not intend it; and can only retrieve my error by reminding him—

> " That the world is always turning on its axis;
> Mankind turns with it, whether heads or tails:
> We live and die, make love, and pay our taxes,
> And as the veering wind shifts, shift our sails."

**THE END.**

LONDON:
Printed by Day & Son, Limited,
Gate Street, Lincoln's Inn Fields.